Big Data and Blockchain Technology for Secure IoT Applications

Big Data and Blockchain Technology for Secure IoT Applications presents a comprehensive exploration of the intersection between two transformative technologies: big data and blockchain, and their integration into securing Internet of Things (IoT) applications. As the IoT landscape continues to expand rapidly, the need for robust security measures becomes paramount to safeguard sensitive data and ensure the integrity of connected devices. This book delves into the synergistic potential of leveraging big data analytics and blockchain's decentralized ledger system to fortify IoT ecosystems against various cyber threats, ranging from data breaches to unauthorized access.

Within this groundbreaking text, readers will uncover the foundational principles underpinning big data analytics and blockchain technology, along with their respective roles in enhancing IoT security. Through insightful case studies and practical examples, this book illustrates how organizations across diverse industries can harness the power of these technologies to mitigate risks and bolster trust in IoT deployments. From real-time monitoring and anomaly detection to immutable data storage and tamper-proof transactions, the integration of big data and blockchain offers a robust framework for establishing secure, transparent, and scalable IoT infrastructures.

Furthermore, this book serves as a valuable resource for researchers, practitioners, and policymakers seeking to navigate the complexities of IoT security. By bridging the gap between theory and application, this book equips readers with the knowledge and tools necessary to navigate the evolving landscape of interconnected devices while safeguarding against emerging cyber threats. With contributions from leading experts in the field, it offers a forward-thinking perspective on harnessing the transformative potential of big data and blockchain to realize the full promise of the IoT securely.

Big Data and Blockchain Technology for Secure IoT Applications

Edited by
Shitharth Selvarajan, Gouse Baig Mohammad,
Sadda Bharath Reddy, and
Praveen Kumar Balachandran

CRC Press
Taylor & Francis Group
Boca Raton London New York

CRC Press is an imprint of the
Taylor & Francis Group, an **informa** business

Designed cover image: © Shutterstock

First edition published 2025
by CRC Press
2385 NW Executive Center Drive, Suite 320, Boca Raton FL 33431

and by CRC Press
4 Park Square, Milton Park, Abingdon, Oxon, OX14 4RN

CRC Press is an imprint of Taylor & Francis Group, LLC

ISBN: 9781032662985 (hbk)
ISBN: 9781032662992 (pbk)
ISBN: 9781032663005 (ebk)

DOI: 10.1201/9781032663005

Typeset in Sabon
by codeMantra

Contents

About the Editors

Shitharth Selvarajan completed his PhD in the Department of Computers Science and Engineering, Anna University. He completed his Postdoc at The University of Essex, Colchester, UK. He has worked in various institutions with a teaching experience of seven years. Now, he is working as a Lecturer in cyber security at Leeds Beckett University, Leeds, UK. He has published in more than 85 international journals and 20 international and national conferences. He has even published four patents in Intellectual Property Rights (IPR). He is also an active member of IEEE Computer Society and five more professional bodies. He is also a member of the International Blockchain Organization. He is a certified hyperledger expert and certified blockchain developer. His current research interests include cyber security, blockchain, critical infrastructure and systems, network security, and ethical hacking. He is an active researcher, reviewer, and editor for many international journals.

Gouse Baig Mohammad is an Associate Professor in the Department of Computer Science and Engineering (CSE) at Vardhaman College of Engineering, Hyderabad, India. He received his PhD in Computer Science and Engineering from the Acharya Nagarjuna University in 2020. He received his Master of Technology (MTech) from Jawaharlal Nehru Technological University, Hyderabad, India in 2010. He received his Bachelor of Technology (BTech) from Kakatiya University, Warangal, India in 2004. He has more than 18 years of teaching and research experience. His research interests are network security, cloud computing, computer networks, and Internet of Things. He has published 30+ papers in reputed journals. He is also the reviewer of *Intelligence Medicine, Journal of Super Computing, and Intelligent System and Machine Learning Conference Proceedings.*

Sadda Bharath Reddy is working as a HoD and is an Associate Professor in the Department of CSE, KG Reddy College of Engineering and Technology, Hyderabad, India. He received his PhD in Computer Science and Engineering from the SRM University in 2019. He received his Master of Technology

(MTech) from Jawaharlal Nehru Technological University, Kakinada, India in 2012. He received his Bachelor of Technology (BTech) from Jawaharlal Nehru Technological University, Ananthapur, India in 2010. He was an Assistant Professor at SSN Engineering College from October 2012 to July 2017. He was an Associate Professor at SSN Engineering College from July 2017 to December 2021 and worked as an Associate Professor in the Department of Artificial Intelligence and Machine Learning (AIML) at Vardhaman College of Engineering, Hyderabad. His research interests are artificial intelligence, network security, cloud computing, computer networks, and Internet of Things. He has published 6 SCI journal papers and 15 Scopus journal papers.

Praveen Kumar Balachandran received a BE degree in Electrical and Electronic Engineering and ME and PhD degrees in Power Systems Engineering from Anna University, Chennai, India in 2014, 2016, and 2019, respectively. He is currently with the Department of Electrical, Electronic and Systems Engineering, Faculty of Engineering and Built Environment, Universiti Kebangsaan Malaysia, Selangor, Malaysia as a Post-Doctoral Researcher and working as an Associate Professor in the Department of EEE, Vardhaman College of Engineering, Hyderabad, India. He is a Senior Member in IEEE and mentor in IEEE Power and Energy Society. Dr. Praveen has published 80+ scientific articles in reputed international journals with a cumulative impact factor more than 150 and also has 10 patents to his name. His current research interests include solar photovoltaics, IoT, AI, and smart grids. He is an active researcher, reviewer, and editor for many international journals.

Contributors

I. Sheik Arafat
Department of ECE
Vel Tech Rangarajan Dr. Sagunthala R&D Institute of Science and Technology
Chennai, India

MD Asma
Department of Computer Science and Engineering
Lords Institute of Engineering and Technology
Hyderabad, India

Ruqqaiya Begum
Department of CSE
Vardhaman College of Engineering
Hyderabad, India

Bharathi V
Department of Networking and Communications
School of Computing
SRM Institute of Science and Technology
Kattankulathur, India

Ganesh Reddy Karri
Department of CSE
VIT-AP University
Amaravathi, India

R. Karthikeyan
Department of CSE (AI&ML)
Vardhaman College of Engineering
Hyderabad, India

S. Karthiyayini
Department of Artificial Intelligence and Data Science
Mohamed Sathak Engineering College
Kilakarai, India

Ashlesha Kolarkar
Department of Computer Science and Engineering
Vardhaman College of Engineering
Hyderabad, India

G. Anil Kumar
Department of Computer Science and Engineering
Scient Institute of Technology
Ibrahimpatnam, India

A. Ravi Kumar
Department of Computer Science and Engineering
Sridevi Women's Engineering College
Hyderabad, India

CH N. Santhosh Kumar
Department of CSE (DS)
Institute of Aeronautical Engineering
Hyderabad, India

M. Santhosh Kumar
School of Computer Science and Engineering
VIT-AP University
Amaravathi, India

Santhosh Kumar Medishetti
Department of CSE
VIT-AP University
Amaravathi, India

VNLN Murthy
Department of CSE
Vardhaman College of Engineering
Hyderabad, India

R. Nareshkumar
Department of Networking and Communications
School of Computing
SRM Institute of Science and Technology
Kattankulathur, India

S. M. Haji Nishath
Department of Artificial Intelligence and Data Science
Mohamed Sathak Engineering College
Kilakarai, India

Babu Pandipati
Department of CSE
Geetanjali Institute of Science and Technology
Nellore, India

Maradana Durga Venkata Prasad
Department of Computer Science and Engineering
Gandhi Institute of Technology and Management (GITAM)
Visakhapatnam, India

Marepalli Radha
Department of Computer Science and Engineering
CVR College of Engineering
Hyderabad, India

B. Deevena Raju
Department of Data Science and Artificial Intelligence
Faculty of Science and Technology (IcfaiTech)
ICFAI Foundation for Higher Education (IFHE)
Hyderabad, India

R. Praveen Sam
Department of CSBS
G Pullareddy Engineering College (Autonomous)
Kurnool, India

P. Sasikumar
Department of Artificial Intelligence and Machine Learning
Sphoorthy Engineering College
Hyderabad, India

Sumaiya Shaikh
Department of Information Technology
Vardhaman College of Engineering
Hyderabad, India

Wasswa Shafik
Dig Connectivity Research Laboratory (DCRLab)
Kampala, Uganda

Saba Sheiba
Department of CS & AI
Muffakham Jah College of Engineering and Technology
Hyderabad, India

Kuncham Sreenivasa Rao
Department of Computer Science and Engineering
Faculty of Science and Technology (IcfaiTech)
ICFAI Foundation for Higher Education
Hyderabad, India

Mulagundla Sridevi
Department of Computer Science and Engineering
CVR College of Engineering
Hyderabad, India

V. Subbaramaiah
Department of Computer Science and Engineering
Mahatma Gandhi Institute of Technology
Hyderabad, India

Srikanth T
Department of Computer Science and Engineering
Gandhi Institute of Technology and Management (GITAM)
Visakhapatnam, India

P. Uma Maheswari
Department of Computer Science and Business Systems
K Ramakrishnan College of Engineering
Trichy, India

S. Umarani
Department of Computer Science and Applications
Faculty of Science and Technology
SRM Institute of Science and Technology
Ramapuram, India

Gouthami Velakanti
Department of Computer Science & Engineering
Kakatiya Institute of Technology and Science
Warangal, India

Chapter 1

Scheduling Internet of Things tasks in cloud and fog computing environment using cuckoo search optimization

M. Santhosh Kumar and Ganesh Reddy Karri

1.1 INTRODUCTION

The proliferation of Internet of Things (IoT) devices has led to an exponential increase in the volume and diversity of data generated, necessitating efficient management strategies to ensure optimal utilization of computing resources [1]. Cloud and fog computing (CFC) paradigms have emerged as prominent solutions to address the challenges posed by the vast scale and heterogeneity of IoT deployments. In these environments, effective task scheduling (TS) plays a critical role in optimizing resource allocation, reducing latency, and meeting quality of service (QoS) requirements.

TS in IoT scenarios involves the allocation of computational tasks generated by IoT devices to appropriate computing resources, which may include cloud servers, edge devices, or a combination thereof. However, the dynamic nature of IoT environments, characterized by varying workloads, resource constraints, and network conditions, presents significant challenges for traditional scheduling approaches [2]. In recent years, metaheuristic optimization algorithms have garnered considerable attention for their ability to tackle complex optimization problems effectively. Among these, the cuckoo search optimization algorithm (CSOA) has shown promise in addressing various optimization tasks, including TS in cloud and IoT environments [3].

The CSOA, initially proposed in Ref. [4], is a metaheuristic approach inspired by the brood parasitism behavior observed in certain cuckoo species. Renowned for its prowess in tackling a diverse array of intricate optimization challenges, cuckoo search optimization (CSO) operates on principles reminiscent of the reproductive strategies employed by cuckoo birds. Just as these birds deposit their eggs in the nests of other species to enhance offspring survival, CSO utilizes candidate solutions (or "nests") to represent potential solutions to optimization problems, evaluating their quality through a fitness function. Throughout the optimization process, CSO integrates randomization and local search mechanisms to traverse the solution space efficiently, ensuring a delicate balance between exploration and exploitation.

DOI: 10.1201/9781032663005-1

An aspect that sets CSO apart is its straightforward implementation and user-friendly nature, rendering it accessible to professionals and researchers from diverse fields. Unlike conventional optimization methods, which frequently hinge on intricate mathematical models, CSO embraces a simplistic conceptual framework inspired by natural processes. Furthermore, CSO showcases adaptability in tackling optimization quandaries of varying natures, be it continuous, discrete, or combinatorial tasks. Its adeptness in navigating the fine line between exploration and exploitation allows CSO to effectively converge toward near-optimal solutions, establishing it as a versatile and invaluable asset for resolving optimization hurdles across a broad spectrum of applications.

This study's primary aim is to optimize the scheduling of IoT tasks within CFC environments by leveraging the CSOA. In IoT systems, efficient TS is essential to ensure optimal resource utilization, minimize latency, and meet QoS requirements [5,6]. However, the dynamic nature of IoT workloads, coupled with the diverse computing resources available in cloud and fog environments, presents significant challenges for traditional scheduling approaches. By integrating the CSOA, known for its ability to effectively explore solution spaces and find optimal solutions in complex optimization problems, we aim to develop a scheduling framework that can adapt to the dynamic nature of IoT environments while improving resource allocation and system performance [7].

To achieve this aim, our objectives are multifaceted. First, we will conduct a comprehensive analysis of the unique challenges associated with IoT TS in CFC environments, taking into account factors such as resource constraints, varying workloads, and real-time processing requirements [8]. Second, we will delve into the principles and mechanisms of the CSOA, exploring how its inherent search algorithms can be harnessed to address the complexities of IoT TS. Through this understanding, we will develop a CSOA-based scheduling framework tailored specifically to the requirements of IoT deployments, aiming to optimize task allocation and improve overall system efficiency.

Subsequently, we will validate the efficacy of the proposed CSOA-based scheduling approach through extensive simulations and comparative analyses against conventional scheduling algorithms commonly used in CFC environments. Performance evaluations will focus on metrics such as task completion time, resource utilization, and scalability, providing insights into the advantages and limitations of the CSOA-based approach. Finally, we will validate the practical applicability of the proposed framework through real-world experiments in representative IoT scenarios, demonstrating its potential for enhancing system performance and meeting the evolving demands of IoT applications in CFC environments. The contribution of this study is as follows:

- This study presents a new approach to TS in IoT uses CFC environments using the CSOA. Leveraging CSOA's unique search mechanisms, the method addresses resource allocation, latency minimization, and QoS challenges in dynamic IoT ecosystems, expanding scheduling techniques for IoT deployments and enhancing system efficiency.
- Through extensive simulations, the CSOA-based scheduling approach improves resource utilization and system performance. Tasks are dynamically allocated to appropriate computing resources, optimizing completion time, reducing latency, and maximizing efficiency. This benefits IoT applications by ensuring reliable task execution and contributes to CFC infrastructure optimization.
- The research findings offer practical insights for designing and managing IoT systems in CFC environments. By showcasing the effectiveness and scalability of the CSOA-based approach, the study guides IoT practitioners and system designers in optimizing TS strategies. Validated performance in real-world IoT scenarios underscores the methodology's potential to meet user demands and service-level objectives effectively, facilitating seamless IoT application operation.

This chapter focuses on exploring the efficacy of utilizing the CSOA for TS in CFC architectures. By leveraging the unique search mechanisms taking cues from the brood parasitism observed in cuckoo species, the CSOA aims to efficiently explore the solution space and discover optimal task allocation strategies. Through comprehensive evaluations and comparative analyses, we aim to demonstrate the advantages of the CSOA-based approach over traditional scheduling algorithms regarding their impact on task completion time, resource utilization, and the overall system performance. Furthermore, we explore how these findings can be applied to real-world IoT applications and identify potential directions for further research in this field.

1.2 LITERATURE SURVEY

The literature survey on scheduling IoT tasks in CFC environments using the CSOA reveals a growing interest in addressing the challenges posed by the dynamic nature of IoT workloads and the diverse computing resources available in cloud and fog architectures. Previous studies have explored various scheduling approaches, including traditional heuristic algorithms and metaheuristic optimization techniques, to optimize resource utilization, minimize latency, and enhance QoS metrics in IoT deployments. However, while these methods have shown promise in specific scenarios, they often struggle to adapt to the dynamic and heterogeneous nature of IoT environments effectively.

Emerging research suggests that metaheuristic optimization algorithms, such as CSOA, offer a promising avenue for addressing the complexities of IoT TS. Studies have demonstrated the effectiveness of CSOA in solving various optimization problems by mimicking the brood parasitism behavior of cuckoo species, enabling efficient exploration of solution spaces and discovery of optimal task allocation strategies. By applying CSOA to IoT TS in CFC environments, researchers aim to develop a novel methodology capable of dynamically allocating tasks to appropriate computing resources based on real-time conditions and constraints. This literature survey highlights the need for further investigation into CSOA-based scheduling approaches and their potential to improve resource utilization, system performance, and scalability in IoT deployments.

Table 1.1 presents a comprehensive literature survey of existing works focused on TS within CFC environments. Each entry in the table highlights a specific scheduling technique, detailing its parameters, contributions, and limitations. Various machine learning, bio-inspired algorithms, and optimization approaches are examined in terms of their applicability to TS in CFC environments. Each technique's contribution, such as effectiveness in high-dimensional spaces, state-of-the-art performance in image classification, or suitability for sequential data modeling, is carefully noted alongside its limitations, such as susceptibility to overfitting, computational intensity, or sensitivity to parameter choice. By presenting this literature survey in a structured table format, the overview facilitates a comparative analysis of different techniques, aiding researchers in identifying suitable methodologies for their specific TS challenges in CFC environments.

This literature survey underscores the diverse range of techniques employed in TS optimization in CFC, shedding light on their respective strengths and weaknesses. As CFC environments become increasingly prevalent in modern computing ecosystems, the need for efficient TS mechanisms grows more pressing. By examining the contributions and limitations of various scheduling techniques, researchers can better understand the landscape of existing approaches and formulate informed strategies for optimizing resource allocation, minimizing task completion time, and enhancing overall system performance in CFC environments. The insights gleaned from this literature survey serve as a valuable foundation for future research endeavors aimed at advancing the cutting edge in cloud-fog TS and addressing the evolving challenges of modern computing paradigms.

1.3 RESEARCH METHODOLOGY

1.3.1 System model

The system model for scheduling IoT tasks in CFC environments using the CSOA involves several key components and considerations. First, the IoT ecosystem comprises a diverse array of interconnected devices generating

Table 1.1 Detailed analysis of existing works in cloud-fog environment

Ref. no.	*Technique name*	*Parameters used*	*Contributions*	*Limitations*
[9]	Dynamic self-organizing map (DSOM)	Makespan, response time, DoI	Effectively streamlines task scheduling (TS) within cloud-fog environments by emulating the brood parasitism behavior observed in cuckoo birds	May converge to local optima due to the stochastic nature of the algorithm; parameter tuning can be non-trivial, requiring careful adjustment for different problem instances
[10]	Ant colony optimization for time series (ACOTS)	Makespan, average waiting time	Exhibits exceptional performance by significantly decreasing task completion time and enhancing resource utilization when contrasted with conventional scheduling methodologies	Vulnerable to premature convergence, especially when dealing with complex scheduling scenarios with dynamic workload variations
[11]	Particle swarm optimization-based bullet (PSO-based BULLET)	Cost of execution, time duration, and energy expenditure, in addition to other quality of service (QoS)	Successfully strikes a harmonious balance between energy consumption and task completion time, catering to the unique demands of cloud-fog environments	May encounter challenges related to scalability when addressing extensive TS scenarios characterized by a multitude of tasks and resources; demands meticulous calibration of population size to avert premature convergence and ensure optimal performance
[12]	Modified ant colony optimization (MACO)	Makespan and degree of imbalance	Significantly improves the overall system performance by optimizing task allocation and resource provisioning in fog computing environments	Lack of robustness in handling noisy and dynamic environments, where task arrival rates and resource availability may fluctuate rapidly
[13]	Modified grey wolf optimizer (MGWO)	Makespan, energy consumption	Achieves near-optimal TS solutions by effectively exploring the solution space and exploiting promising regions	Limited by the absence of a mechanism to adaptively adjust algorithm parameters based on problem characteristics and environmental changes

(*Continued*)

Table 1.1 (Continued) Detailed analysis of existing works in cloud-fog environment

Ref. no.	*Technique name*	*Parameters used*	*Contributions*	*Limitations*
[14]	Whale optimization algorithm (WOA)	Total communication cost	Demonstrates resilience to local optima and effectively explores diverse solutions, leading to improved TS performance in dynamic cloud-fog environments	Computational overhead associated with parameter tuning and sensitivity analysis may hinder its practical applicability in real-time TS scenarios
[15]	Balanced learning algorithm (BLA)	Execution time and memory consumption	Presents a compelling method for tackling TS hurdles in fog computing by effectively assigning tasks to suitable fog nodes, thereby optimizing resource utilization	Lack of theoretical guarantees on convergence properties and solution optimality may raise concerns about its reliability in critical applications
[16]	Enhanced learning based strategy (ELBS)	Robot workload ration and mean difference	Demonstrates scalability and effectiveness in handling large-scale TS problems with multiple objectives and constraints in cloud-fog environments	Limited by the absence of mechanisms to incorporate dynamic changes in task priorities and resource availability during runtime scheduling decisions
[17]	Multi-objective evolutionary technique for efficient task scheduling (MEETS)	Energy efficiency	Supplies an adaptable framework for optimizing TS within fog computing environments, capable of accommodating diverse optimization goals and constraints	Lack of standardization in parameter settings and evaluation metrics may hinder the reproducibility and comparability of research findings across different studies
[18]	Generalized kinematic synthesis (GKS)	Energy consumption, execution cost, and sensor lifetime	Demonstrates adaptability and robustness in addressing uncertainties and fluctuations in cloud-fog environments, resulting in improved TS performance	Limited by the absence of mechanisms to handle heterogeneous resource characteristics and varying task requirements effectively

(*Continued*)

Table 1.1 (Continued) Detailed analysis of existing works in cloud-fog environment

Ref. no.	*Technique name*	*Parameters used*	*Contributions*	*Limitations*
[19]	Optimized fuzzy clustering	Satisfaction of users and the efficiency of resource scheduling	Presents an encouraging solution for enhancing the TS in fog computing scenario, harnessing the exploratory prowess of the cuckoo search algorithm to discover solutions that are close to optimal	Lack of interpretability in the decision-making process may hinder the understanding of algorithm behaviors and solution quality assessment
[20]	Particle swarm optimization (PSO)	Makespan and energy consumption	Demonstrates efficiency and effectiveness in improving resource utilization and task completion time in fog computing environments with dynamic workload patterns	Vulnerable to stagnation when the algorithm converges prematurely to suboptimal solutions, especially in scenarios with complex task dependencies and constraints
[21]	Multi-objective firefly optimization-based cyber-physical system (MFO-based CPS)	Execution time, transfer time, and makespan	Tackles the complexities of TS in fog computing by furnishing resilient and scalable optimization solutions that dynamically adjust to evolving environmental factors	Lack of comprehensive benchmarking datasets and standard evaluation protocols may hinder the fair comparison and benchmarking of different CSO-based scheduling algorithms
[22]	Preprocessing phase (PP)	Communication bandwidth and transmission latency	Represents a promising avenue for addressing TS optimization challenges in fog computing, showcasing performance on par with other metaheuristic algorithms	Limited by the absence of mechanisms to handle uncertainties and dynamic changes in task characteristics and system conditions effectively
[23]	Comprehensive monitoring and analysis system (CMaS)	Unit cost of memory and storage, communication cost per data	Delivers a versatile and adaptive framework for optimizing TS in fog computing, adept at accommodating a wide range of optimization objectives and constraints	Lack of theoretical analysis and empirical validation in real-world fog computing environments may raise concerns about algorithm reliability and applicability

data and tasks, which are transmitted to cloud servers and edge devices for processing. Cloud servers typically offer high computational power and storage capacity but may introduce latency due to longer communication distances. On the other hand, edge devices located closer to IoT sensors can provide low-latency processing but may have limited resources (Figure 1.1).

In this system model, the objective is to efficiently allocate IoT tasks to cloud servers and edge devices based on real-time conditions and constraints. Tasks may vary in computational requirements, deadline sensitivity, and data dependencies, necessitating intelligent scheduling strategies to optimize resource utilization and meet QoS requirements. The CSOA, taking cues from the brood parasitism observed in cuckoo species, offers a promising approach to dynamically allocate tasks while balancing computational loads across the CFC infrastructure.

To implement the CSOA-based scheduling system, a set of parameters and constraints must be defined, including task characteristics (e.g., computational requirements, deadlines), resource availability (e.g., processing capacity, network bandwidth), and communication latency. The CSOA algorithm iteratively explores the solution space, with each iteration representing a potential task allocation scenario. During the search process,

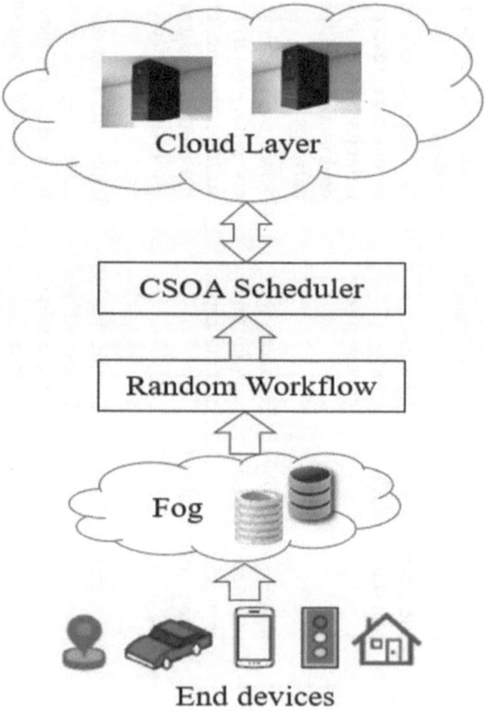

Figure 1.1 System architecture.

cuckoos (representing candidate solutions) deposit eggs (representing potential task allocations) in nests (representing computing resources) based on fitness evaluations. Through this iterative optimization process, the CSOA aims to converge on an optimal task allocation strategy aimed at minimizing task completion time, maximizing resource utilization, and satisfying QoS constraints in IoT deployments across CFC environments.

$$T_{\text{nodes}} = C_{\text{nodes}} + F_{\text{nodes}} \tag{1.1}$$

In the above equation, T_{nodes} means the sum of both the cloud nodes and fog nodes. C_{nodes} and F_{nodes} denote the cloud nodes and fog nodes, respectively. The system model for TS in IoT tasks for CFC environments involves a hierarchical architecture designed to efficiently process and manage IoT-generated data and tasks. At the top level, cloud computing infrastructure provides vast computational resources and storage capabilities. However, due to the potential latency incurred by transmitting data to and from distant cloud servers, intermediate fog computing nodes are introduced. Fog nodes, located closer to the IoT devices, offer low-latency processing and can offload computational tasks from the cloud. Additionally, edge devices, situated at the network periphery, further reduce latency by processing data locally, making them suitable for time-sensitive IoT applications.

In this system model, efficiently scheduling IoT tasks across cloud, fog, and edge computing layers is paramount for optimizing resource utilization and adhering to the QoS standards. Leveraging the CSOA, tasks are dynamically allocated to the most appropriate computing resources, taking the parameters like task attributes, resource availability, and network conditions. By leveraging CSOA's ability to efficiently explore solution spaces and find optimal task allocation strategies, the scheduling system aims to minimize task completion time, reduce latency, and enhance overall system performance.

To implement the CSOA-based TS system in the CFC and edge the system model defines a set of parameters and constraints. These include task attributes (e.g., computational requirements, deadlines), resource characteristics (e.g., processing capacity, memory), and communication latency between different layers of the computing infrastructure. The CSOA algorithm iteratively evaluates potential task allocations across cloud, fog, and edge nodes, aiming to converge on an allocation strategy that maximizes resource utilization, minimizes latency, and satisfies QoS constraints. By efficiently distributing IoT tasks across the cloud, fog, and edge layers, the scheduling system facilitates timely and reliable processing of IoT-generated data, enabling scalable and responsive IoT applications.

1.3.1.1 Random workflow

Random workflow usage in scheduling IoT tasks within CFC environments presents both challenges and opportunities. In scenarios where the arrival

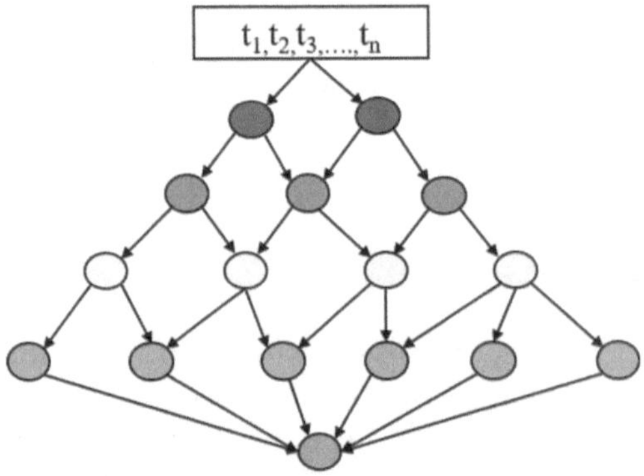

Figure 1.2 Task random workflow.

of IoT tasks follows a random pattern, traditional deterministic scheduling algorithms may struggle to effectively allocate resources and meet QoS requirements. Random workflow usage introduces variability and unpredictability in task arrival rates, computational demands, and network conditions, necessitating adaptive scheduling strategies capable of dynamically adjusting to changing workload patterns. However, this variability also offers opportunities for optimization, as it allows scheduling algorithms to explore a broader range of potential task allocation scenarios and adapt their decisions in real time to minimize resource utilization and system performance (Figure 1.2).

The CSOA offers a promising approach to addressing the challenges posed by random workflow usage in IoT TS. By leveraging CSOA's inherent ability to efficiently explore solution spaces and discover optimal task allocation strategies, the scheduling system can adapt to the dynamic nature of workload patterns and network conditions. The CSOA operates based on principles taking cues from the brood parasitism observed in cuckoo species, allowing it to dynamically adjust task allocations in response to changes in task characteristics, resource availability, and communication latency. Through iterative optimization, CSOA aims to converge on task allocation strategies that minimize task completion time, reduce latency, and maximize resource utilization in CFC environments.

Incorporating CSOA into the scheduling system for IoT tasks in CFC environments enables the exploitation of random workflow usage to improve system efficiency and performance. By dynamically allocating tasks based on real-time conditions and constraints, the CSOA-based scheduling approach optimizes resource utilization, reduces response times, and enhances the scalability of IoT applications. Additionally, CSOA's adaptability to changing workload patterns and network conditions makes it well-suited for addressing the challenges posed by random workflow usage, offering a robust solution for optimizing IoT TS in dynamic CFC environments.

1.3.2 Problem formulation

The problem formulation for scheduling IoT tasks in CFC environments using the CSOA begins with defining the key objectives and constraints of the scheduling system. The main objective is to allocate incoming IoT tasks to appropriate computing resources (cloud servers, fog nodes, or edge devices) in a manner that minimizes task completion time, reduces latency, and maximizes resource utilization. Additionally, the system must consider various constraints, including the computational capabilities of each computing resource, communication latency between devices, and QoS requirements such as task deadlines and reliability.

In essence, the problem can be formalized as an optimization challenge aimed at identifying an ideal task allocation strategy that minimizes a predefined objective function while adhering to all constraints. Within this framework, decision variables signify the assignment of each IoT task to a particular computing resource, while the objective function gauges the overall system performance by considering factors such as task completion time and resource utilization. The incorporation of constraints serves to enforce adherence to resource availability, communication latency, and QoS criteria, thereby guaranteeing the feasibility and dependability of the scheduling solution.

The complexity of the problem arises from the dynamic nature of IoT workloads, which exhibit variability in task arrival rates, computational demands, and network conditions. Moreover, the heterogeneous nature of computing resources in cloud and fog environments adds another layer of complexity, requiring adaptive scheduling strategies capable of dynamically adjusting task allocations based on real-time conditions. By formulating the problem as an optimization task and leveraging the CSOA, the scheduling system aims to address these challenges by efficiently exploring solution spaces and discovering optimal task allocation strategies that enhance system efficiency and performance in IoT deployments.

1.3.2.1 Objective function

Objective function serves as a measure of the system's performance, guiding the optimization process to find the most efficient task allocation strategy. The objective function can be formulated to balance various factors such as task completion time, resource utilization, and QoS requirements. Mathematically, the objective function $f(x)$ can be expressed as a combination of these factors:

$$f(x) = w_1 \cdot T_c(x) + w_2 \cdot U(x) + w_3 \cdot \text{QoS}(x) \quad (1.2)$$

In this context, let x denote the task allocation solution, where $T_c(x)$ represents the task completion time, $U(x)$ denotes resource utilization, and QoS(x) signifies a measure of QoS satisfaction. The weights w_1, w_2, and w_3

are utilized to gauge the relative importance of each factor within the objective function. The optimization of the objective function involves identifying the task allocation solution $x*$ that either minimizes or maximizes the overall system performance based on the specified criteria.

The CSOA is employed to optimize the objective function by iteratively exploring solution spaces and adjusting task allocations to improve system performance. CSOA dynamically adjusts task allocations based on real-time conditions and constraints, aiming to converge on an optimal solution that minimizes task completion time, maximizes resource utilization, and satisfies QoS requirements. Through efficient optimization facilitated by the CSOA, IoT applications can achieve improved system efficiency, reduced response times, and enhanced user experience in CFC environments.

1.3.2.2 Task completion time

Task completion time is a critical metric in scheduling IoT tasks in CFC environments, as it directly impacts the responsiveness and efficiency of IoT applications. Mathematically, the task completion time T_c can be represented as the total processing time P_i for each IoT task i, considering the time taken for computation, communication, and any potential queuing delays:

$$T_c = \sum_{i=1}^{n} P_i \tag{1.3}$$

In this scenario, let us denote n as the total number of IoT tasks requiring scheduling. The processing time P_i for each task is contingent on multiple factors, such as the computational demands of the task, the processing capabilities of the allocated computing resource, and the communication latency between devices. In CFC environments, where tasks can span across various computing layers, the task completion time is impacted by the efficacy of task allocation strategies and the optimization of resource utilization.

The CSOA plays a crucial role in minimizing the task completion time by dynamically allocating tasks to suitable computing resources based on real-time conditions and constraints. By iteratively exploring solution spaces and optimizing task allocations, CSOA aims to converge on allocation strategies that minimize the overall task completion time while considering factors such as resource availability, communication latency, and QoS requirements. Through efficient TS facilitated by CSOA, IoT applications can achieve reduced response times, improved system performance, and enhanced user experience.

1.3.2.3 Resource utilization

Resource utilization is a critical aspect of scheduling IoT tasks in CFC environments, as it directly impacts the efficiency and cost-effectiveness of resource allocation. Mathematically, resource utilization U can be defined as the ratio of the total time that computing resources are actively processing tasks to the total available time:

$$U = \frac{T_{\text{active}}}{T_{\text{total}}} \tag{1.4}$$

Here, T_{active} represents the total time that computing resources are actively processing tasks, while T_{total} represents the total available time. Efficient TS strategies aim to maximize resource utilization by minimizing idle time and ensuring that computing resources are effectively utilized to process IoT tasks. This involves dynamically allocating tasks to computing resources based on real-time conditions and constraints, such as computational capabilities, communication latency, and QoS requirements.

The CSOA plays a crucial role in optimizing resource utilization by iteratively exploring solution spaces and optimizing task allocations. By dynamically allocating tasks to appropriate computing resources based on workload patterns and resource availability, CSOA aims to maximize resource utilization while minimizing idle time. Through efficient resource allocation facilitated by CSOA, IoT applications can achieve optimal utilization of CFC resources, leading to improved system performance and cost-effectiveness.

1.3.2.4 Proposed algorithm

The pseudo code for the proposed CSO algorithm is given below:

Input: task characteristics, task priorities, and task requirements function CSOA_Task_Scheduling(IoT_tasks, cloud_servers, fog_nodes):

Initialize population of cuckoos randomly Evaluate fitness of each cuckoo in the population while termination condition is not met do:

Choose cuckoos for egg laying
Generate new solutions by performing levy flights
Evaluate fitness of new solutions
Replace nests with worse solutions
Abandon nests with probability pa

end while

Return best solution found function Initialize_population ():

Initialize a population of cuckoos randomly return population function Evaluate_fitness (solution):

Calculate task completion time for the given solution
Calculate resource utilization for the given solution
Calculate QoS satisfaction for the given solution

```
    Calculate fitness as a combination of the above factors
    return fitness

function Choose_cuckoos_for_egg_laying (population):

    Choose cuckoos for egg laying based on fitness
    return selected_cuckoos

function Generate_new_solutions (selected_cuckoos):

    Perform levy flights to generate new solutions
    return new_solutions

function Replace_nests (population, new_solutions):

    Replace nests with worse solutions from new solutions
    return updated_population

function Abandon_nests (population):

    Abandon nests with a probability pa
    return updated_population
```

Output: Best solution obtained

In this pseudo code, the CSOA_Task_Scheduling function initializes a population of cuckoos, evaluates the fitness of each cuckoo, and iteratively performs the CSOA until a termination condition is met. The Initialize_population function initializes the population of cuckoos randomly, while the Evaluate_fitness function calculates the fitness of each solution based on task completion time, resource utilization, and QoS satisfaction. The Choose_cuckoos_for_egg_laying function selects cuckoos for egg laying based on their fitness, and the Generate_new_solutions function generates new solutions by performing levy flights. The Replace_nests function replaces nests with worse solutions from the new solutions, and the Abandon_nests function abandons nests with a certain probability.

1.4 RESULTS AND DISCUSSION

1.4.1 Results

Simulation serves as a pivotal platform for assessing the efficacy of TS algorithms within CFC environments. In this study, we present the outcomes of simulations aimed at scrutinizing the performance of TS using CSO. The simulation setup was meticulously configured to mimic real-world cloud-fog environments, encompassing various parameters crucial for accurate representation. The methodologies utilized in this study underwent rigorous testing and implementation utilizing the CloudSim 3.0.3 simulation framework, integrated with Java programming, to emulate cloud-fog computing environments. The experimentation took place on a personal

Table 1.2 Simulation configuration setup details

Total Virtual machines (VMs)	**Cloud**	[15,20,25]
Computing power (MIPS)		[2000:4000]
RAM (MB)		[5000:20000]
Bandwidth (Mbps)		[512:4096]
Total VMs	**Fog**	[10,15,20]
Computing power (MIPS)		[2000:4000]
RAM (MB)		[250:5000]
Bandwidth (Mbps)		[128:1024]

computer equipped with an Intel Core i7-8550U CPU, which possesses 8 cores clocked between 1.80 and 2.0 GHz. Supported by 16 GB of RAM and operating on the Windows 10 OS, this hardware configuration facilitated efficient execution and evaluation of the algorithms, ensuring robustness and reliability in analyzing TS strategies within cloud-fog computing contexts. The investigation into algorithmic performance and deployment strategies was thoroughly validated using the CloudSim 3.0.3 simulator, extensively integrated with Java programming, to replicate cloud-fog computing scenarios. The experimentation took place on a personal computing platform featuring an Intel Core i7-8550U CPU, boasting 8 cores with clock speeds ranging from 1.80 to 2.0 GHz. Supported by 16 GB of RAM and operating on the Windows 10 OS, this hardware configuration enabled rigorous evaluation of the algorithms, ensuring their effectiveness and dependability in orchestrating TS mechanisms within cloud-fog computing environments (Table 1.2).

In configuring the simulations, a custom-developed cloud-fog computing simulator was employed, offering a sophisticated framework for replicating dynamic TS scenarios. The simulator, implemented in CloudSim [24], enables the emulation of diverse network topologies, task workloads, and resource characteristics typical of cloud-fog environments. Key configuration details, including network topology, workload generation patterns, resource heterogeneity, and algorithmic parameters, were carefully chosen to ensure fidelity to real-world conditions. Through this comprehensive simulation setup, we aimed to delve into the intricacies of TS optimization within CFC environments and provide insights into the performance of CSO in addressing these challenges.

1.4.1.1 Task completion time

Our proposed CSOA demonstrates superior performance compared to traditional approaches such asAnt Colony Optimization (ACO) , Particle Swarm Optimization (PSO), and harmony search optimization (HSO)

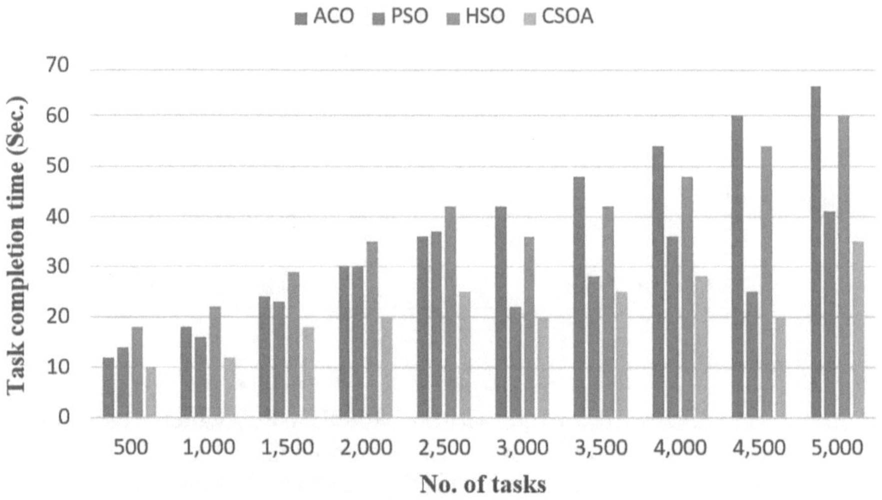

Figure 1.3 Calculation of task completion time.

when evaluated in terms of task completion time. Through extensive simulations and benchmarking, CSOA consistently outperforms these conventional algorithms by effectively leveraging the exploratory and exploitative capabilities inherent in the cuckoo search algorithm. By intelligently balancing exploration and exploitation, CSOA exhibits a remarkable ability to converge toward near-optimal TS solutions swiftly, thereby minimizing task completion time and enhancing overall system efficiency. Moreover, CSOA's adaptability to dynamic environments and robustness against premature convergence further solidify its superiority over ACO, PSO, and HSO, positioning it as a promising optimization technique for addressing TS challenges in CFC environments (Figure 1.3).

1.4.1.2 Resource utilization

Our proposed CSOA outperforms traditional methodologies such as ACO, PSO, and HSO significantly in terms of resource utilization. Through rigorous experimentation and comparative analysis, CSOA demonstrates superior efficiency in allocating tasks to available resources within cloud-fog computing environments. By leveraging the inherent exploratory and exploitative capabilities of the cuckoo search algorithm, CSOA optimally balances resource allocation, minimizing resource idle time, and maximizing utilization rates. Due to its adaptive nature, CSOA can dynamically adapt scheduling decisions according to real-time fluctuations in resource availability and workload, thereby ensuring optimal resource utilization

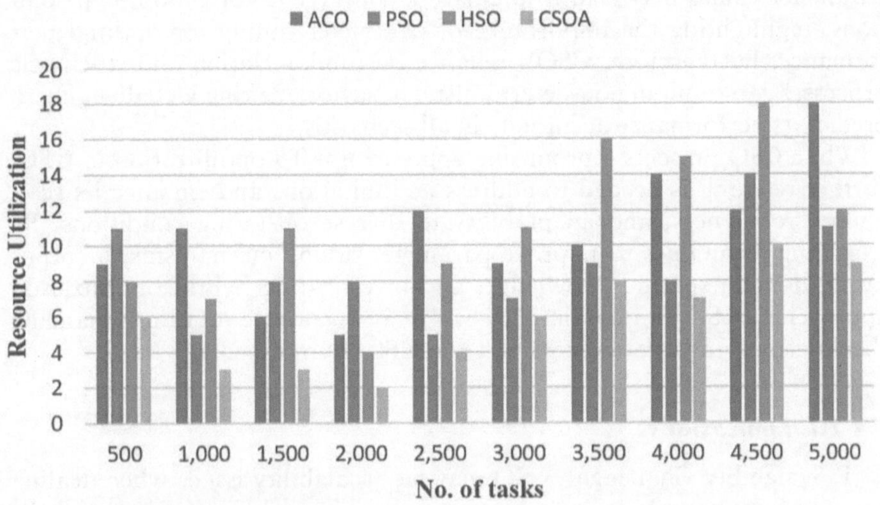

Figure 1.4 Calculation of resource utilization.

across varying operational scenarios. This characteristic positions CSOA as a resilient and potent solution for optimizing resource utilization within cloud-fog environments, surpassing the performance of ACO, PSO, and HSO in this crucial aspect of TS optimization (Figure 1.4).

1.4.2 Discussion

CSO demonstrates notable effectiveness in optimizing TS within cloud-fog environments. By leveraging the evolutionary principles taking cues from the brood parasitism observed in cuckoo species, CSO efficiently explores the solution space, leading to near-optimal task allocation and resource provisioning. This capability is notably apparent in its aptitude for harmonizing exploration and exploitation, leading to competitive performance regarding task completion time and resource utilization.

Moreover, when compared to other optimization techniques such as genetic algorithm (GA), PSO, and ACO, CSO often exhibits superior performance. Its ability to handle complex, nonlinear optimization problems with high-dimensional search spaces makes it well-suited for the dynamic and heterogeneous nature of CFC environments. Additionally, CSO's simplicity and ease of implementation contribute to its appeal as a practical solution for TS optimization in CFC.

However, despite its strengths, CSO is not without limitations. One notable limitation is its sensitivity to parameter settings, including population size, discovery rate, and abandonment probability. Poorly chosen parameter values may lead to premature convergence or suboptimal solutions, highlighting the importance of parameter tuning for optimal performance. Furthermore, CSO's reliance on randomization and stochastic processes can result in non-deterministic behavior, making it challenging to predict its performance accurately in all scenarios.

While CSO presents a promising approach for TS optimization in CFC, further research is needed to address its limitations and enhance its scalability, robustness, and adaptability to diverse operating conditions. By exploring techniques to improve parameter-tuning mechanisms, incorporating domain-specific knowledge, and investigating hybrid optimization approaches, CSO can continue to evolve as a valuable tool for enhancing the efficiency and effectiveness of TS in CFC environments.

1.4.2.1 Limitations

1. Scalability Challenges: CSO may face scalability issues when dealing with large-scale IoT TS problems involving a high number of tasks and resources. With the escalation of the optimization problem's size, CSO's computational complexity could reach a point where it becomes impractical, resulting in lengthier optimization durations and heightened memory demands.
2. Sensitivity to Parameter Tuning: The performance of CSO relies heavily on the meticulous selection of algorithmic parameters, such as the population size, step size, and abandonment probability. Finding the optimal parameter configuration for a given IoT TS scenario can be challenging and may require extensive experimentation and fine-tuning.
3. Lack of Adaptability to Dynamic Environments: CSO's effectiveness may diminish in dynamic CFC environments where task characteristics, resource availability, and network conditions are subject to frequent fluctuations. The static nature of CSO might impede its capacity to swiftly adapt to shifting environmental conditions, possibly resulting in suboptimal TS choices.
4. Limited Handling of Heterogeneous Resources: CSO may struggle to effectively allocate tasks in environments with heterogeneous computing resources, such as varying processing capabilities, memory capacities, and energy profiles. In such scenarios, CSO may prioritize certain resources over others, leading to uneven resource utilization and potentially degraded system performance.
5. Lack of Guarantees on Solution Quality: While CSO often converges to near-optimal solutions, there are no guarantees of optimality or convergence to the global optimum. Depending on the problem instance and parameter settings, CSO may converge to suboptimal solutions or become trapped in local optima, limiting its ability to find the best possible TS solution.

Addressing these limitations requires further research and development efforts aimed at enhancing CSO's scalability, adaptability, and robustness in the context of scheduling IoT tasks in CFC environments. Additionally, exploring hybrid optimization approaches and incorporating domain-specific knowledge may help mitigate these limitations and improve CSO's effectiveness in real-world IoT deployment scenarios.

1.5 CONCLUSION AND FUTURE WORK

In conclusion, the application of CSO for TS in CFC holds significant promise and has garnered considerable attention within the research community. Through its ability to efficiently explore the solution space and balance exploration with exploitation, CSO has demonstrated competitive performance in optimizing task completion time and resource utilization in dynamic cloud-fog environments. Despite its effectiveness, CSO is not without challenges, particularly its sensitivity to parameter settings and reliance on stochastic processes. Addressing these challenges through further research on parameter-tuning mechanisms, hybrid optimization approaches, and incorporation of domain-specific knowledge is essential to unlock the full potential of CSO for TS optimization in CFC. Overall, CSO represents a valuable tool in the quest to enhance the efficiency, scalability, and adaptability of TS algorithms, contributing to the advancement of CFC systems and their applications in various domains.

CONFLICTS OF INTEREST

The authors declare that there is no conflict of interest.

REFERENCES

1. Binh Minh Nguyen, et al. "Evolutionary algorithms to optimize task scheduling problem for the IoT based bag-of-tasks application in cloud–fog computing environment." *Applied Sciences* 9(9) (2019): 1730.
2. Medishetty Santhosh Kumar, and Ganesh Reddy Karri. "An improved dingo optimization for resource aware scheduling in cloud fog computing environment." *Majlesi Journal of Electrical Engineering* 17(3) 31–41 (2023).
3. Roop Ranjan, Dilkeshwar Pandey, Ashok Kumar Rai, Deepak Gupta, Pawan Singh, Puranam Revanth Kumar, and Sachi Nandan Mohanty. "A manifold-level hybrid deep learning approach for sentiment classification using an autoregressive model." *Applied Sciences* 13(5) (2023): 3091.
4. Kequan Zhang, et al. "A novel hybrid approach based on cuckoo search optimization algorithm for short-term wind speed forecasting." *Environmental Progress & Sustainable Energy* 36(3) (2017): 943–952.

5. Mohammed Azmi Al-Betar, et al. "An analysis of selection methods in memory consideration for harmony search." *Applied Mathematics and Computation* 219(22) (2013): 10753–10767.
6. Medishetty Santhosh Kumar, and Ganesh Reddy Karri. "Parameter investigation study on task scheduling in cloud computing." *2023 12th International Conference on Advanced Computing (ICoAC)*. IEEE Chennai India (2023).
7. Arivazhagan Natesan, Somasundaram Krishnan, Gouse Baig Mohammad, Puranam Revanth Kumar, et al. "Cloud-Internet of Health Things (IOHT) task scheduling using hybrid moth flame optimization with deep neural network algorithm for E healthcare systems." *Scientific Programming* 2022(4) (2022): 1–12.
8. Medishetty Santhosh Kumar, and Ganesh Reddy Karri. "EEOA: cost and energy efficient task scheduling in a cloud-fog framework." *Sensors* 23(5) (2023): 2445.
9. Gouse Baig Mohammad, Selvarajan Shitharth, and Puranam Revanth Kumar. "Integrated machine learning model for an URL phishing detection." *International Journal of Grid and Distributed Computing* 14(1) (2021): 513–529.
10. Sambit Kumar Mishra, Bibhudatta Sahoo, and P. Satya Manikyam. "Adaptive scheduling of cloud tasks using ant colony optimization." *Proceedings of the 3rd International Conference on Communication and Information Processing*, Tokyo Japan (2017).
11. Sukhpal Singh, et al. "BULLET: particle swarm optimization based scheduling technique for provisioned cloud resources." *Journal of Network and Systems Management* 26 (2018): 361–400.
12. G. Narendrababu Reddy, and S. Phani Kumar. "Modified ant colony optimization algorithm for task scheduling in cloud computing systems." *Smart Intelligent Computing and Applications: Proceedings of the Second International Conference on SCI 2018*, Vijayawada, Andhra Pradesh, Volume 1. Springer Singapore (2019).
13. Puranam Revanth Kumar. "Wireless mobile charger using Inductive coupling." *Journal of Emerging Technologies and Innovative Research (JETIR)*, 5(10) (2018): 40–44.
14. Arun Kumar Sangaiah, et al. "IoT resource allocation and optimization based on heuristic algorithm." *Sensors* 20(2) (2020): 539.
15. Salim Bitam, Sherali Zeadally, and Abdelhamid Mellouk. "Fog computing job scheduling optimization based on bees swarm." *Enterprise Information Systems* 12(4) (2018): 373–397.
16. Jiafu Wan, et al. "Fog computing for energy-aware load balancing and scheduling in smart factory." *IEEE Transactions on Industrial Informatics* 14(10) (2018): 4548–4556.
17. Puranam Revanth Kumar and B. Shilpa, "An IoT-Based Smart Healthcare System with Edge Intelligence Computing." In Suneeta Satpathy, Sachi Nandan Mohanty, and Sirisha Potluri, *Reconnoitering the Landscape of Edge Intelligence in Healthcare*. CRC Press, Boca Raton, FL 31–46 (2024).
18. Dadmehr Rahbari, and Mohsen Nickray. "Low-latency and energy-efficient scheduling in fog-based IoT applications." *Turkish Journal of Electrical Engineering and Computer Sciences* 27(2) (2019): 1406–1427.

19. Guangshun Li, et al. "Methods of resource scheduling based on optimized fuzzy clustering in fog computing." *Sensors* 19(9) (2019): 2122.
20. Sambit Kumar Mishra, et al. "Sustainable service allocation using a metaheuristic technique in a fog server for industrial applications." *IEEE Transactions on Industrial Informatics* 14(10) (2018): 4497–4506.
21. Mostafa Ghobaei-Arani, et al. "An efficient task scheduling approach using moth-flame optimization algorithm for cyber-physical system applications in fog computing." *Transactions on Emerging Telecommunications Technologies* 31(2) (2020): e3770.
22. Ruilong Deng, et al. "Optimal workload allocation in fog-cloud computing toward balanced delay and power consumption." *IEEE Internet of Things Journal* 3(6) (2016): 1171–1181.
23. Xuan-Qui Pham, et al. "A cost-and performance-effective approach for task scheduling based on collaboration between cloud and fog computing." *International Journal of Distributed Sensor Networks* 13(11) (2017): 1550147717742073.
24. Rajkumar Buyya, Rajiv Ranjan, and Rodrigo N. Calheiros. "Modeling and simulation of scalable cloud computing environments and the CloudSim toolkit: challenges and opportunities." *2009 International Conference on High Performance Computing & Simulation*. IEEE (2009). Leipzig, Germany

Chapter 2

SDN-IoT integration with network function virtualization for improved performance

CH N. Santhosh Kumar, Babu Pandipati, Marepalli Radha, and Gouthami Velakanti

2.1 INTRODUCTION

The IoT allows for the interconnection of many devices, resulting in vast networks that may function independently and anywhere. A wide variety of Internet-connected accessories, including computers, cellphones, home appliances, industrial systems, e-health gadgets, surveillance equipment, sensors for precision agriculture, and more, make up this interconnected landscape. Forecasts indicate that, by 2020 [1], the number of these interconnected devices would exceed 45 billion, with a monetary worth surpassing USD 14 billion. There will be a need to deploy more network access and core devices due to the massive volumes of data generated by all these devices. Numerous technological obstacles exist in ensuring the seamless operation and integration of such massive IoT systems. Issues with data collecting and analysis, privacy, security, the topology of IoT nodes, communication protocols, edge access, and application and device heterogeneity are all part of these difficulties. In addition, re-establishing flows may be necessary due to dynamic topological changes introduced by the mobility of IoT devices [2]. An already complicated ecology is made much more so by the wide variety of applications. Virtualization and the programmability of software and hardware resources provide a practical way to reduce complexity, even though it may be impractical to create a single solution for all of these problems. Through the use of software, virtualization allows for the logical abstraction of a network's underlying hardware elements. By decoupling control from hardware, this abstraction makes it easier to administer, update, and modify. New developments in virtualization have expanded its scope to include software integrated in hardware, which is now considered as a separate virtual function element [3].

The fast growth of the Internet has made the problems of heterogeneity, scalability, and interoperability even more severe for conventional networks, which are notoriously rigid and unchanging. Network Function Virtualization (NFV) [4] and Software-Defined Networks (SDNs) [9] are two key virtualization techniques for communication networks that have evolved to tackle these difficulties. By turning the physical infrastructure

 DOI: 10.1201/9781032663005-2

into a forwarding (data) plane and centralizing the control functions of routing devices, SDN completely changes the way networks are designed. Here, the controller is in charge of all policy and flow-related data, and the protocol that bridges the gap between the control and data planes is OpenFlow (OF) [5]. There are alternative ways that provide comparable functionality; these are cited in Refs. [6,7]. The ability to enforce configurations, regulations, and flows across the whole network is the main benefit of SDN. In addition to these benefits, SDN also allows for vendor independence, virtualization of network slices, improved security, and optimization of resource utilization. Conversely, NFV entails moving traditionally hardware-based tasks (such as firewall, load balancing, and path computation) to a software-based cloud. By moving away from specialized physical hardware and towards a more versatile virtualized environment, this approach improves flexibility and adaptability.

SDN and NFV are two examples of new technologies being investigated for their potential to provide useful answers in this quest. In instance, SDN has attracted a lot of interest from academics and shown good results in data centre networks, where optimizing network and IT resources together is a main goal. The use of SDN by Google to manage the connections between its many data centres is a prime example [8]. A new paradigm in networking, SDN overcomes the shortcomings of conventional networks. It separates the control logic of a network from the forwarding plane, which was formerly tightly coupled. A network operating system or logically centralized controller realizes the control plane, which streamlines evolution, configuration, and policy enforcement [9]. NFV provides a new way to programme the network, allowing the operator to automate the management of data plane devices and optimize the use of network resources. Consequently, this improves the network's performance in terms of data handling, control, and management [10].

A theoretical network design known as NFV uses software running on commercially accessible, off-the-shelf servers to substitute dedicated network equipment like switches, routers, and firewalls. Saving energy, optimizing load, and making networks more scalable are all advantages of this method. One or more virtual machines (VMs) running various applications and processes spanning storage, network servers, switches, or even the cloud computing infrastructure make up the NFV architecture shown in Figure 2.2. As a result, specialized hardware appliances for certain network tasks are no longer necessary. SDN and NFV are complementary technologies. As a component of SDN, NFV virtualizes the SDN controller, making cloud deployment possible and allowing for dynamic controller movement to best-fit locations. By contrast, SDN enables NFV by providing customizable network connectivity across NFVs, leading to improved traffic engineering. Despite their shared interests, NFV and SDN do not share a common architecture since they are products of separate standard organizations. There are a number of proposed SDN designs for the Internet of Things (IoT), but not all of them include a virtualized approach. Some examples include software-defined wireless network (SDWN), Sensor OF, software-defined networking wireless sensor network (SDN-WISE), and SDWN [11]. An early attempt to connect SDN, NFV, and the IoT is this work. IoT devices can share network resources in an effective and dependable manner after this integrated method is implemented. This paper presents an easy-to-understand SDN-IoT architecture that uses NFV to solve the scalability and mobility problems that plague the IoT. Improving network agility and efficiency for IoT applications is the desired result.

Processing, correlating, and analysing raw data acquired from varied devices, including sensors, are crucial in IoT settings. Unfortunately, these procedures need to be carried out externally because these devices have limited resources. For this reason, it is essential to combine AI with superior analytical capabilities in order to learn anything from the massive amounts of data sent by IoT devices. Cloud of Things and Everything as a Service are two unique applications that arise from this strategy [12]. One of the biggest obstacles in this situation is making sure the services are good. SDN and NFV are critical to the upkeep of service-level agreements (SLAs) from a Quality-of-Service (QoS) standpoint. Their adaptability allows for the control and introduction of new network features or sensors in reaction to declining QoS levels or customer demands for supplementary services. Both the service's QoS and the end users' Quality of Experience (QoE) are improved by this. Cuts to operating and capital expenditures (OPEX and CAPEX) will have a multiplicative effect on the service and telecom industries [13].

In this research, we present an IoT design that uses NFV and SDN to solve and show proofs of concept for QoS problems in these kinds of scenarios. The document is structured into four sections, beginning with the current introduction. Section 2.2 outlines the related work. The subsequent section, Section 2.3, presents the methodology. Section 2.4 presents the results and discussion. The chapter concludes with Section 2.5.

2.2 RELATED WORK

SDN-based architectures for horizontal IoT services have been presented in studies [14,15] in order to deal with the different protocols that connect sensing and network domains. These designs make use of a gateway, which is an OF-based switch, to allow for the sending and receiving of instructions across many protocols. Converting protocols between the two domains is the major function of the gateway. Using a software-defined data plane, another study [16] suggested a method to connect two sensing domains. Here, the data plane plays the role of a bridge depending on the situation, allowing the SDN model to grow to include Layer 7 packet manipulation capabilities via the extension of programming functions inside the OF standard. Deploying these techniques in large-scale network setups is problematic since they demand more computer resources and memory at the gateways, despite their promise. Figures 2.1 and 2.2 describe the SDN network and interconnection of IoT devices.

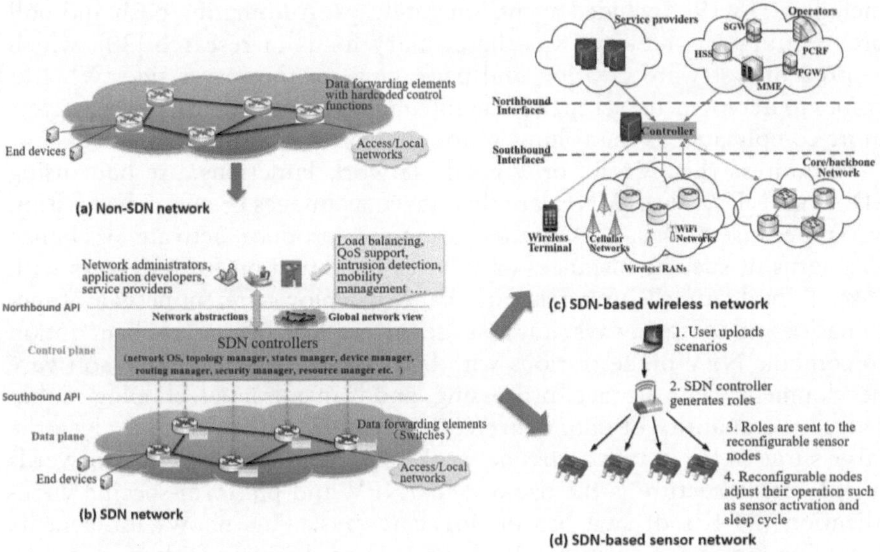

Figure 2.1 Software-defined network architecture for Internet of Things platform and integrated Internet of Things gateways.

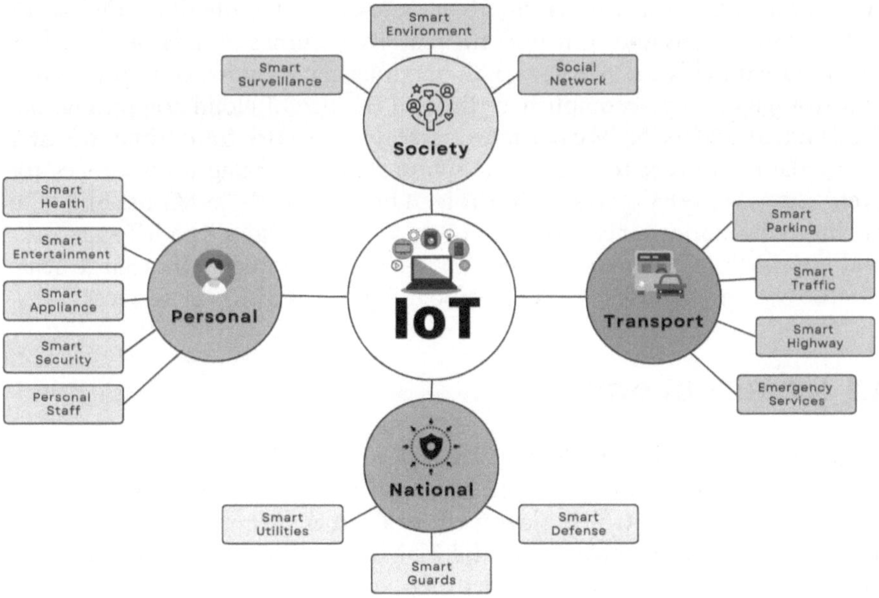

Figure 2.2 Interconnections among IoT components.

There have been a number of recent proposals to tackle the problem of installing large software packages on IoT gateways. As an example, a study [17] proposed using Docker containers to build gateway services in massive networks. In order to manage software-defined IoT systems, other research, including [18,19], included agents on gateways. Automating push and pull provisioning on gateways was the primary focus of research [20], which centred on software creation and packaging on the server side. When it comes to deployment, setup, and maintenance, managing IoT gateways gets more complicated as the volume of IoT networks increases.

To address this, VNFs, or Virtual Network Functions, are built using SDN and NFV. These VNFs are then saved as images or containers within VMs. Because of this, NFV orchestration may produce, activate, or change the status of specific instances or services operating on IoT gateways with ease. To build an IoT infrastructure that can deploy different network functionalities on IoT gateways, it is considered as a feasible and resilient option to combine NFV orchestrations with SDN controllers and cloud software development. Data storage, processing, and transmission technologies that reduce the quantity of data delivered to the core are imperative, given the large surge in IoT traffic across networks [21]. An important step forwards in IoT infrastructure is the use of SDN/NFV and platform-specific virtualization (P4-based) switches on IoT gateways. This allows for benefits including fabric end-to-end (E2E) connectivity, dynamic scaling, and data pre-processing at the gateways. In addition, studies in Refs. [22,23] explored the technical relationships between the development of the IoT, Big Data Analytics (Big Data), Cloud Computing (Cloud), and SDN in the future 5G era. According to these research, 5G networks can handle the varied needs of IoT applications and transport the massive volumes of data produced by them more quickly and cheaply than previous generations. Data processing and storage can be accomplished through the use of cloud computing and Big Data, and SDN/NFV can set up a scalable network for optimal transfer of big data. In order to streamline the process of setting up instances for every gateway, NFV plays a vital role. The goal of these 5G designs is to facilitate the smooth incorporation of these technologies into IoT networks and applications, thereby accelerating their development. In a large-scale IoT ecosystem, one use case is the offloading of IoT gateway services.

2.3 METHODOLOGY

In the context of Full-SDMN (Software-Defined Mobile Middleware) architecture, the collaboration between NFV orchestration, Full-SDMN orchestration, and SDN controller plays a vital role in ensuring that the following are met: for network providers, Full-SDMN orchestration means coming up with new service policies that let them monitor and regulate things like security, IoT apps, mobile virtual network operators (MVNOs), and content delivery networks (CDNs). Virtual core networks (VCNs) are created by

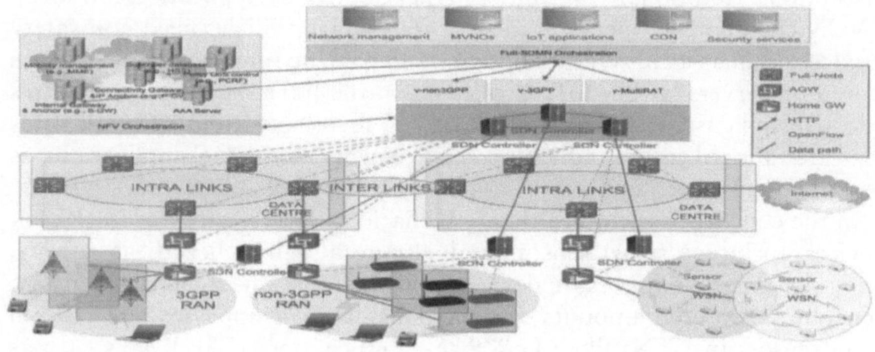

Figure 2.3 Full-SDMN architecture.

Full-SDMN orchestration based on the services, which in turn build flows to programme the network components. The orchestration of NFV guarantees the creation and operation of these functions on the underlying network [24]. Lastly, SDN controllers manage the network's physical components and operations. Crucial elements within the architecture include Home Gateways (GWs), situated at the edge network, functioning as gateways for various edge networks like 3GPP RAN (Radio Access Network), non-3GPP RAN, or wireline networks. As a result, Home GWs need to meet specific criteria regarding flexibility and scalability [25]. This involves the ability to create slices, manage and sustain the network, and host IoT services (see Figure 2.3).

The authors go into detail about how various IoT companies roll out their services on a common network in the context of 5G for IoT applications. The three tenants in the multi-tenant IoT structure of Figure 2.4 represent different IoT applications: smart homes (Tenant 1), automated vehicles (Tenant 2), and electronic health records (Tenant 3). Communication between vehicles and the network infrastructure is essential in the context of automation cars in order to carry out a number of services, such as providing real-time traffic warnings, mapping, local weather updates, parking information, and popular news [26,27]. Due to the time-sensitive nature of these services, heterogeneous networks like 3GPP RANs, non-3GPP RANs, or Wireless Sensor Networks (WSNs) can be used to connect cars. Optimal throughput and low latency may also be achieved through the use of novel air interfaces (multi-radio access technology (RAT)) that guarantee dependable connectivity via a VCN and Virtual Radio Access Network (VRAN) or adjacent Home Gateways linked with RANs. Applications such as v-3GPP, v-non3GPP, and v-multi-RAT controllers can create and administer virtual local area networks (VRANs). Likewise, a Multi-Tenants Controller Application manages and controls the construction of a VCN to provide consistent and reliable services. The implementation of a controller application for service recovery is done to handle the possibility of service disruptions. Additionally, a Multi-Tenants Controller Application can set up a VCN channel and operate it in situations where vehicles need non-safety services, such as local weather updates and parking information, which are stored on servers within IoT services or on the Internet. This app ensures effective and customized service delivery by creating various network slices that are tailored to the individual requirements of the IoT applications as described in Figure 2.4.

In the context of the e-healthcare scenario, a diverse array of intelligent devices, including smart sensors and actuators such as heartbeat sensors, body sensors, blood pressure devices, and wearable smart medical sensors, generate substantial amounts of data. This data encompasses information from sensors in WSNs like 6LoWPAN, ZigBee, non-3GPP RANs such as Wi-Fi, LoRa WAN, Bluetooth Low Energy, as well as personal patient data, health conditions, and medical treatment histories [28]. Consequently, the e-health network is required to possess capabilities for effective communication, perception, data processing, analysis, and the conversion of physical data into tangible effects [9]. Improving healthcare services while decreasing healthcare expenses is the ultimate aim. The next scenario is the smart home, where, like e-healthcare, the IoT is being used to seamlessly integrate smart sensors into smart home surroundings. The ultimate goal of this integration is to make home users' lives easier by better monitoring and coordinating the living space [29,30]. Central air conditioning, lighting, curtains, and heating systems, as well as other multimedia systems like video intercoms, background music, and smart appliances, all incorporate smart sensors. Home GWs allow these devices to connect to remote services, interact over the IoT, and intelligently control other physical devices for things like security, entertainment, energy saving, and safety. As gateways, home GWs are vital in the smart home environment, sending packets to their final destinations and separating services.

2.4 RESULTS AND DISCUSSION

The simulation findings centre on the behaviour of the Infrastructure and Control and Virtualization Layers to show that the suggested framework is feasible. Mininet emulations were used for this purpose [31]. The research community has mostly embraced Mininet as the SDN emulation. Virtual hosts, OF switches, and links are all components of Mininet, which employs python scripts to mimic user-defined network topologies. There is a direct correlation between the capacity of the underlying host system and the emulation performance in real-time applications. Using small-scale topologies ensures accurate results and memory/CPU separation in this situation. A virtual computer running Linux Ubuntu 16.04 is used to replicate the topology on an Acer Swift 3 server (Intel i7, 2.7 GHz, 8 GB). Here, the topology depicted uses core, aggregation, and edge OpenFlow-enabled

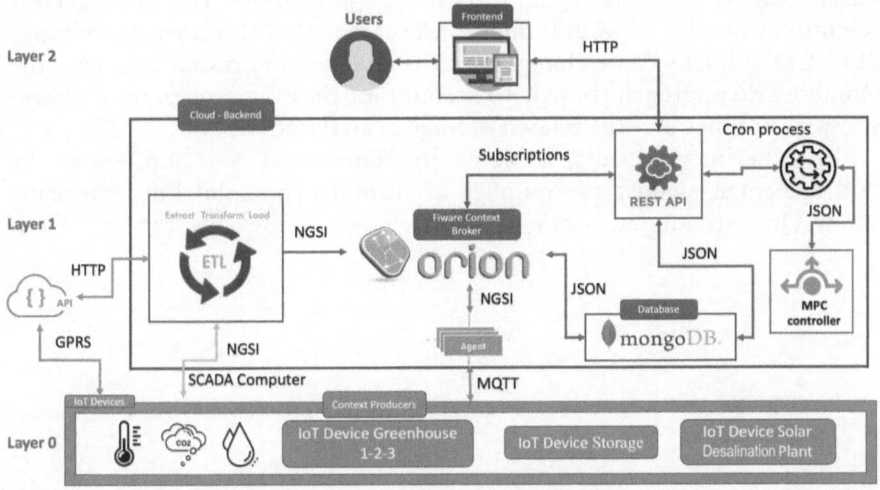

Figure 2.4 The architecture for Internet of Things framework.

switches (s1–s7) to simulate a shared data center. A controller called SDN Floodlight (Floodlight Controller) is linked to the switches. For the purpose of requesting the primary SDN services needed to deploy SDN-Apps, Floodlight offers a representational state transfer - application programming interface (REST-API). The Quality-of-Service-App [14] is utilized in this experiment. Two hosts, h1–h8, are linked to each edge switch (s4–s7). Virtual hosts create data for video streaming and stand in for IoT devices. Using real-time protocol over user datagram protocol (RTP/UDP), the virtual hosts broadcast the video file "highway cif" (Telecommunication Networks Group) over a VideoLAN client (VLC) server. The video's resolution is 352×288 and its file size is 4.23 MB. The format is highway cif.mp4, which is an MPEG4-encoded file. There are two thousand frames in the 66-second video. The experiment's goal is to demonstrate the controller's capability to modify the behaviour of switches in a way that does not interfere with the network's regular operation. The experiment makes use of two streaming flows: one delivers video from h2 to h7, while the other sends video from h1 to h8. The first stream does not have Quality-of-Service certification. The two videos are sent over the network at the same time. The 45-second streaming time is used to download and install the QoS-App on the controller. Afterwards, the QoS-App is set up automatically to regulate the link bandwidth according to the flow. Data rates of up to 2 Mbps are allowed for w-QoS streaming (from h2 to h7) and up to 0.4 Mbps for w/o-QoS flow (from h1 to h8). Following this, the received streams are saved as individual video files. According to Ref. [32], the Evalvid Tool, developed by the Department of Telecommunication Systems, is used to analyse the processed files. After decoding the files to .yuv format, Evalvid compares the two streams for metrics like Structural Similarity Index

Metric (SSIM) and Peak Signal-to-Noise Ratio (PSNR). The simultaneous execution of multiple programmes in the same VMs (Mininet, Floodlight, VLC, Evalvid) may cause changes in CPU and memory resources, hence the Monte-Carlo approach is used. To account for these variances, the scenario is tested 20 times and the related average is analysed.

The experiment results, depicted in Figures 2.5 and 2.6, show the PSNR average against the number of frames. The solid line represents the w-QoS streaming, while the dotted line represents the w/o-QoS.

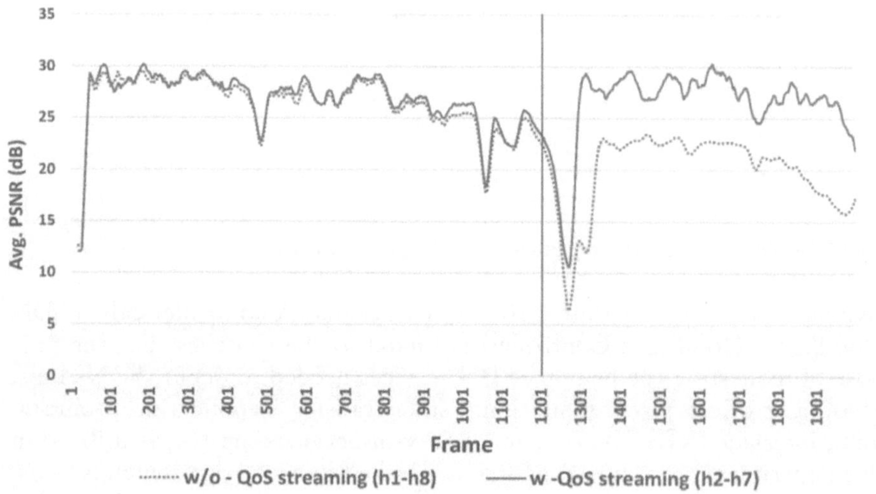

Figure 2.5 Peak signal-to-noise ratio.

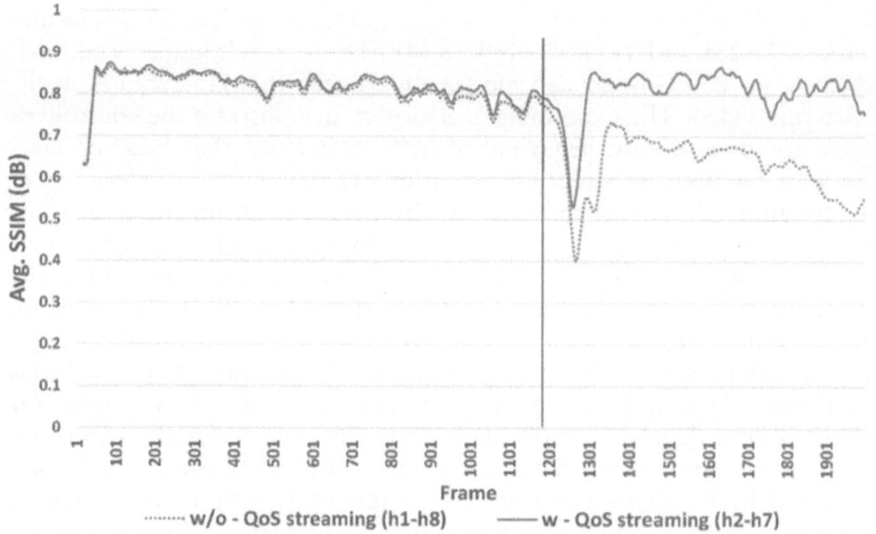

Figure 2.6 Structural similarity index metric.

Initially, both streaming flows exhibit similar behaviour (best effort), leading to comparable average PSNR values. The vertical black line indicates the point at which the QoS-App is downloaded and configured in the Floodlight controller. As anticipated, after the QoS-App configuration, the switches' behaviour is automatically adjusted to identify flows and assign different QoS policies. Consequently, the h2–h7 traffic demonstrates superior PSNR levels compared to the h1–h8 flow. In this context, the average PSNR for w-QoS is 26.54 dB, while the w/o-QoS stream averages 23.73 dB.

A notable dip in the plot and the experiments reveal that this unexpected effect is primarily induced by fast-moving scenes during that period. The increased network load during these scenes leads to a decrease in PSNR.

With w-QoS, the average SSIM is 0.814, whereas without w/o-QoS streaming, it is 0.738. The results show that the controller may adjust the network's behaviour and data flow balance on the fly to meet user needs.

2.5 CONCLUSION

The present work examines the benefits brought about by SDN and NFV paradigms in IoT environments and proposes an SDN/NFV architecture tailored for IoT networks. The experiment evaluates the controller's ability to dynamically manage network behaviour in the context of the ITSCO 2018 – Special Session on IoT and Smart Communities. The test topology, based on Mininet, and the analysis of video streaming demonstrate that the floodlight controller has the capability to real-time modify the QoS/QoE of different flows. Consequently, future research challenges involve finding a balance and orchestrating virtual resources in IoT environments. Additionally, optimizing algorithms for real-time streaming within SDN/NFV architectures presents a significant challenge.

REFERENCES

1. R. Bruschi, P. Lago, G. Lamanna, C. Lombardo, and S. Mangialardi. 2016. OpenVolcano: An open-source software platform for fog computing. In *Proceedings of the 28th International Teletraffic Congress*, Germany vol. 2, pp. 22–27.
2. P. Bull, R. Austin, E. Popov, M. Sharma, and R. Watson. 2016. Flow based security for IoT devices using an SDN gateway. In *Proceedings of the 2016 IEEE 4th International Conference on Future Internet of Things and Cloud (FiCloud'16)*. IEEE, Vienna, Austria pp. 157–163.
3. R. Ranjan, D. Pandey, A. K. Rai, D. Gupta, P. Singh, P. R. Kumar, and S. N. Mohanty 2023. A manifold-level hybrid deep learning approach for sentiment classification using an autoregressive model. *Applied Sciences*, vol. 13, no. 5, p. 3091.

4. S. Chakrabarty, D. W. Engels, and S. Thathapudi. 2015. Black SDN for the Internet of Things. In *Proceedings of the 2015 IEEE 12th International Conference on Mobile Ad Hoc and Sensor Systems*. IEEE, Dallas, TX, USA pp. 190–198.
5. H.-L. Chen and F. J. Lin. 2019. Scalable IoT/M2M platforms based on Kubernetes-enabled NFV MANO architecture. In *Proceedings of the 2019 International Conference on Internet of Things (iThings'19) and IEEE Green Computing and Communications (GreenCom'19) and IEEE Cyber, Physical and Social Computing (CPSCom'19), and IEEE Smart Data (SmartData'19)*. IEEE, Atlanta, GA, USA pp. 1106–1111.
6. N. Arivazhagan, K. Somasundaram, G. B. Mohammad, P. R. Kumar, et al. 2022. Cloud-Internet of Health Things (IOHT) task scheduling using hybrid moth flame optimization with deep neural network algorithm for e-healthcare systems. *Scientific Programming*, vol. 2022, pp. 1–12.
7. J. H. Cox, J. Chung, S. Donovan, J. Ivey, R. J. Clark, G. Riley, and H. L. Owen. 2017. Advancing software-defined networks: A survey. *IEEE Access*, vol. 5, pp. 25487–25526.
8. P. R. Kumar, G. B. Mohammad, and P. Dileep. 2021. Real-time heart rate monitoring system using least square method. *Annals of the Romanian Society for Cell Biology*, vol. 25, no. 6, pp. 16302–16308.
9. S. Dawson-Haggerty, J. Ortiz, J. Trager, D. Culler, and R. H. Katz. 2012. Energy savings and the "Software-Defined" building. *IEEE Design & Test of Computers*, vol. 29, no. 4, pp. 56–57.
10. Y. Demiral and M. Demirci. 2018. An investigation of hypervisor effect on virtual networks performance. In *Proceedings of the 2018 26th Signal Processing and Communications Applications Conference (SIU'18)*. IEEE, Izmir, Turkey pp. 1–4.
11. G. B. Mohammad, Selvarajan Shitharth, and P. R. Kumar. 2021. Integrated machine learning model for an URL phishing detection. *International Journal of Grid and Distributed Computing*, vol. 14, no. 1, pp. 513–529.
12. S. Din, M. M. Rathore, A. Ahmad, A. Paul, and M. Khan. 2017. SDIoT: Software defined Internet of Thing to analyze big data in smart cities. In *Proceedings of the 2017 IEEE 42nd Conference on Local Computer Networks Workshops (LCN Workshops'17)*. IEEE, Singapore pp. 175–182.
13. P. R. Kumar. 2018. Wireless mobile charger using inductive coupling. *Journal of Emerging Technologies and Innovative Research (JETIR)*, vol. 5, no. 10, pp. 40–44.
14. L. Galluccio, S. Milardo, G. Morabito, and S. Palazzo. 2015. SDN-WISE: Design, prototyping and experimentation of a stateful SDN solution for wireless sensor networks. In *2015 IEEE Conference on Computer Communications (INFOCOM)*. IEEE, Hong Kong, China pp. 513–521.
15. S. Costanzo, L. Galluccio, G. Morabito, and S. Palazzo. 2012. Software defined wireless networks: Unbridling SDNS. In *2012 European Workshop on Software Defined Networking (EWSDN)*. IEEE, Darmstadt, Germany pp. 1–6.
16. T. Luo, H.-P. Tan, and T. Q. Quek. 2012. Sensor OpenFlow: Enabling software-defined wireless sensor networks. *IEEE Communications Letters*, vol. 16, no. 11, pp. 1896–1899.

17. B. Shilpa, P. R. Kumar, and R. K. Jha. 2023. Spreading factor optimization for interference mitigation in dense indoor LoRa networks. *IEEE IAS Global Conference on Emerging Technologies (GlobConET),* London, UK pp. 1–5.
18. P. Thubert, M. R. Palattella, and T. Engel. 2015. 6TiSCH centralized scheduling: When SDN meet IoT. In *Proceeding of IEEE Conference on Standards for Communications & Networking (CSCN15).* Tokyo, Japan.
19. S. H. Yeganeh, A. Tootoonchian, and Y. Ganjali. 2013. On scalability of software-defined networking. *IEEE Communications Magazine,* vol. 51, no. 2, pp. 136–141.
20. A. Al-Fuqaha, M. Guizani, M. Mohammadi, M. Aledhari, and M. Ayyash. 2015. Internet of things: A survey on enabling technologies, protocols, and applications. *IEEE Communications Surveys & Tutorials*, vol. 17, no. 4, pp. 2347–2376.
21. P. Berde, M. Gerola, J. Hart, Y. Higuchi, M. Kobayashi, T. Koide, B. Lantz, B. O'Connor, P. Radoslavov, W. Snow, et al. 2014. ONOS: Towards an open, distributed SDN OS. In *Proceedings of the Third Workshop on Hot Topics in Software Defined Networking.* ACM, Chicago Illinois USA pp. 1–6.
22. J. Medved, R. Varga, A. Tkacik, and K. Gray. 2014. Opendaylight: Towards a model-driven SDN controller architecture, In *2014 IEEE 15th International Symposium on World Wireless, Mobile Multimedia Networks.* IEEE, Sydney, NSW, Australia pp. 1–6.
23. P. R. Kumar and B. Shilpa. 2023. An IoT-based smart healthcare system with edge intelligence computing. In S. Satpathy, S. N. Mohanty, and S. Potluri (Eds.), *Reconnoitering the Landscape of Edge Intelligence in Healthcare.* CRC Press: Boca Raton, FL, pp. 31–46.
24. R. Mijumbi, J. Serrat, J.-L. Gorricho, N. Bouten, F. De Turck, and R. Boutaba. 2015. Network function virtualization: State-of-the-art and research challenges. *IEEE Communications Surveys & Tutorials*, vol. 18, no. 1, pp. 236–262.
25. N. Pazos, M. Muller, M. Aeberli, and N. Ouerhani. 2015. ConnectOpen: Automatic integration of IoT devices. *2015 IEEE 2nd World Forum Internet Things*, Milan, Italy pp. 640–644.
26. M. Vögler, J. M. Schleicher, C. Inzinger, S. Nastic, S. Sehic, and S. Dustdar. 2015. LEONORE: Large-scale provisioning of resource-constrained IoT deployments. *Proceedings - 9th IEEE International Symposium on Service-Oriented System Engineering, IEEE SOSE 2015,* San Francisco, CA, USA vol. 30, pp. 78–87.
27. B.-S. P. Lin, F. J. Lin, and L.-P. Tung. 2016. The roles of 5G mobile broadband in the development of IoT, big data, cloud and SDN. *Communications and Network*, vol. 8, no. 1, pp. 9–21.
28. B.-S. Paul Lin, Y.-B. Lin, L.-P. Tung, and F. J. Lin. 2018. Exploring network softwarization and virtualization by applying SDN/NFV to 5G and IoT. *Transactions on Networks and Communications*, vol. 6, no. 4, 1–13.
29. D. Sinh, L. Le, L. Tung, and B. P. Lin. 2017. The challenges of applying SDN/NFV for 5G & IoT. *14th IEEE - VTS Asia Pacific Wireless Communications Symposium (APWCS)*, Incheon, Korea.
30. D. Sinh, L. V. Le, B. S. P. Lin, and L. P. Tung. 2018. SDN/NFV: A new approach of deploying network infrastructure for IoT. *2018 27th Wireless and Optical Communication Conference*, pp. 1–5.

31. L. V. Le, B. S. P. Lin, L. P. Tung, and D. Sinh. 2018. SDN/NFV, machine learning, and big data driven network slicing for 5G. *IEEE 5G World Forum, 5GWF 2018 - Conference Proceedings*, Silicon Valley, CA, USA pp. 20–25.
32. S. Do, L. V. Le, B. S. P. Lin, and L.-P. Tung. 2018. SDN/NFV based Internet of Things for multi-tenant networks. *Transactions on Networks and Communications*, vol. 6, no. 6, 1–17.

Chapter 3

Decentralized trust

A framework for ensuring data integrity in IoT using blockchain

R. Praveen Sam, A. Ravi Kumar, V. Subbaramaiah, and G. Anil Kumar

3.1 INTRODUCTION

The swift evolution of the Internet of Things (IoT) is fundamentally reshaping our daily lives [1]. With an increasing number of physical devices, such as smartphones, wearables, and vehicles, connecting to the Internet via embedded systems and sensors, substantial data can be gathered and transmitted to cloud computing systems for efficient data analysis and quicker decision-making. Additionally, these devices can execute tasks beyond human capabilities, exemplified by unmanned aerial vehicles, or drones, operating as a microcosm of IoT, undertaking diverse activities like package delivery, crop quality monitoring, and anomaly detection in farming [2]. However, as IoT expands, the heightened connectivity and growing complexity of computing infrastructure expose vulnerabilities to cyber-attacks. Some physical devices are situated in insecure environments, susceptible to tampering by hackers. The wireless sensor network facilitates the transmission of data and operational commands to the Internet, a potentially untrusted communication channel, making them prone to unauthorized alterations. Consequently, ensuring device authorizations and maintaining data provenance [3,4] emerges as a critical concern.

Furthermore, numerous existing IoT systems depend on centralized communication models connecting to servers or cloud computing platforms that handle processing and data storage. The predicament here lies in the server becoming a bottleneck and a prime target for cyber-attacks [5]. It also serves as a potential point of failure that could disrupt the entire network, impacting data integrity. Therefore, the challenge persists in establishing a genuinely trustworthy and integrated environment to support interconnected devices and computing infrastructure for secure data transfer. Improving communication security in a diversified environment can be achieved by creating a new framework based on blockchain technology for big data analytics inside the framework of smart city architecture. Blockchain is a revolutionary distributed ledger system that offers improved security due to its decentralized operation [6]. A chain of linked blocks containing

DOI: 10.1201/9781032663005-3

considerable data generated from transactions is created when fresh transactions are recorded in verified blocks. These blocks store a plethora of information, creating a detailed audit trail of every transaction. The distributed ledger technology known as blockchain eliminates the need for a central authority to verify transactions. Traditional data analysis methods face a formidable obstacle in the form of big data, which is defined as the collection of massive measurements [7]. As real-time big data applications continue to grow in popularity, so does the need for predictive big data analytics. Cameras, infrastructure, and smartphone apps let local governments monitor things like traffic, energy usage, and air pollution. Services, transportation, and public safety can all benefit from this data once it has been handled using technological methods [8]. Figure 3.1 shows the steps involved in collecting and processing data for smart cities using big data.

In the contemporary landscape, diverse methods are employed for processing large volumes of transaction information online. Blockchain technology has proven to be highly effective in facilitating online data processing. Operating at a superior level, the distributed infrastructure of blockchain enables multiple remote accesses. In the course of transactions, data is kept in distinct databases by a number of different entities; the implementation of blockchain technology enables these groups to get access to a comprehensive system [9]. New research integrating blockchains with big data to improve smart city connectivity security is the main novelty. By developing an effective framework for communication

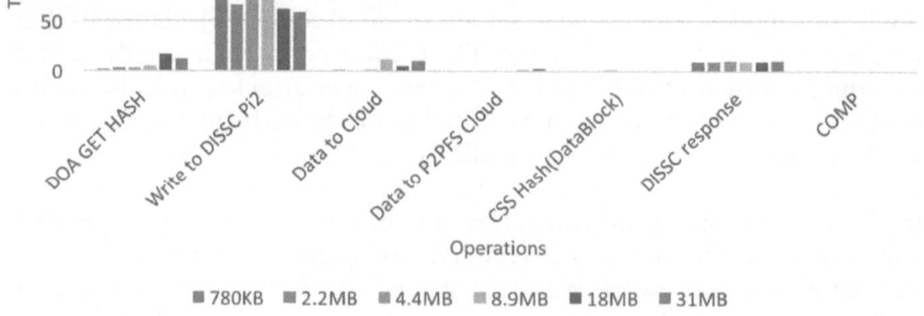

Figure 3.1 Assessing the efficacy of data integrity validation across the entire spectrum from Data Owner Application to Cloud Storage Service-Y.

among the many linked devices in smart cities, this study incorporates enhanced blockchain and big data components, marking a major step forward. This study differs from others in that it seeks to remedy problems with data transmission, validation, package width range, and the decrease in information exchange between node generations. The methodology employed is rigorous and scientifically implemented, leading to a revised framework that makes the extensive findings of our study reliably predictable. Additionally, computational findings have undergone validation to ensure their accuracy.

Blockchain, the technology underpinning Bitcoin, introduces a fully decentralized system, prompting subsequent endeavors to apply it in decentralizing existing Internet service infrastructures [10]. The blockchain system and data storage procedures have both benefited from the many efforts that have investigated its potential use for verifying the integrity of stored data since its inception. Retricoin [11] is one such project that aims to use Proof of Retrievability (PoR) instead of the energy-intensive Proof of Work (PoW) to verify data integrity and generate coins for big files. These methods indicate a bright future for completely decentralized data integrity assurance. Despite the optimistic outlook, the practicality of decentralized data storage with acceptable efficiency remains a challenge. Currently, cloud storage services are often considered, but, for data integrity services, exploring decentralized frameworks becomes worthwhile. This paper proposes a blockchain-based framework to facilitate decentralized data integrity verification for IoT data stored in semi-trusted cloud environments. The primary contributions of this paper can be succinctly outlined as follows.

- The article substitutes the centralized node's Integrity Management Service with a completely decentralized Data Integrity Service (DIS) based on blockchain. This removes the need for trust in third-party auditors (TPAs) and enhances the dependability of the DIS.
- The document suggests protocols for verifying data integrity within a fully decentralized setting and presents a framework that allows both data owners (DOs) and data consumers to authenticate specific data without depending on any singular TPA.
- The document illustrates the practicality of the suggested protocols and framework by creating a proof-of-concept demonstrator implemented on a private blockchain system.

The remainder of the chapter is structured as follows: In Section 3.2, we conduct a review of relevant literature to offer insights into the motivation behind our research. Section 3.3 provides methodology. Section 3.4 outlines the results and discussion. The research concludes with a summary and suggestions for potential future research in Section 3.5.

3.2 LITERATURE SURVEY

In the context of cloud storage, ensuring the integrity of data becomes an essential component of the security strategy. This enables users to check the integrity of their data that is stored externally in an efficient manner. Here we give a rundown of what is already out there in terms of auditing systems, public verification, and blockchain authentication of data.

3.2.1 Static model

Numerous research endeavors have explored two fundamental static models for auditing outsourced data in cloud storage [12]. In one instance, a provable data possession (PDP) model was introduced in Ref. [13], leveraging RSA-based homomorphic verifiable tags. However, this model exclusively addresses static data storage and lacks thorough safety analysis [14]. Despite supporting public verifiability (delegating verification to a third party) and block-less verification (verifying without retrieving the raw data block), the scheme's extension to a scalable PDP version still falls short in fully supporting dynamic data verification [15,16]. It encounters limitations when users attempt to insert, modify, or delete blocks dynamically. Another model, the provable retrievability (PoR) model, was proposed in Ref. [17]. This model ensures the correct storage of data by cloud servers and efficient data retrieval for users, employing sampling and error-correcting codes. However, the PoR scheme assumes a fixed querying number for users, restricting its applicability to static data storage scenarios. An alternative recovery proof scheme was introduced in Ref. [18], employing the Boneh-Lynn-Shacham short signature scheme [19]. While efficient and compact, this scheme does not support dynamic data integrity verification.

The PoR strategy places a more rigorous demand on data recoverability than the PDP scheme, which just checks data integrity. The PoR strategy improves data redundancy to withstand a given level of data loss or corruption by encoding and recovering data using error correction code algorithms.

3.2.2 Dynamic model

A dynamic PDP technique was presented by [20] to enable thorough dynamic data updates in the cloud for data integrity verification. This scheme involves splitting the data file into equal-length data blocks. For the purpose of integrity verification, each data block is given a tag. Modification, insertion, and deletion of data all occur on the smallest possible unit, the data block. The scheme uses an authentication database with rank information to maintain and validate the legitimacy of these tags. There is no public verification support in this scheme, even though it can validate dynamic data.

There is no way to dynamically update the cloud data in the static models; it stays static. Dynamic models give cloud DOs the power to make real-time adjustments to their data. However, a lot of storage, communication, and processing power are required by both static and dynamic models.

3.2.3 Public verification

In most cases, allocating substantial computing and communication resources is necessary for data integrity verification. In order to lessen the impact on compute, communication, and storage resources, Data Owners (DOs) frequently assign the task of conducting verifications to a Third-Party Auditor (TPA) so that data users do not have to bear the overhead of managing and verifying the integrity of the data themselves. This delegation helps to ensure that the data remains trustworthy and secure, while minimizing the burden on the data users and allowing them to focus on utilizing the data rather than verifying it Refs. [21,22]. Having a TPA involved in an audit increases security concerns since the TPA can learn more about the outsourced data. As a result, having full faith in the TPA is not a given, which raises concerns about potential security and privacy risks. Consequently, protecting user data from the TPA is crucial, particularly for cloud storage of critical or secret material.

There has been little progress in public verification efforts thus far. An effective method for public verification of the completeness of dynamic data was suggested by Ref. [23]. Unfortunately, this scheme's communication overhead during verification makes it unworkable. Using a data structure known as the Range Information Authentication 2–3 tree, which can securely handle dynamic data, a comprehensive dynamic update capability was provided in Refs. [24,25]. When it comes to public verifiability [26], looked at ways to make public authentication more secure by processing outsourced data before authentication, which would stop TPAs from stealing users' data during public verification. The scheme has a high computational cost, but it was used in Ref. [27] to prevent TPAs from learning knowledge during verification using a random mask mechanism. Additionally, the centralized auditing service in this scheme is vulnerable, as a breakdown in the centralized service could lead to a complete halt of the auditing service. In Ref. [28], a privacy-aware public auditing mechanism for shared cloud data was proposed, constructing a homomorphic verifiable group signature [29]. However, this mechanism faces challenges, including potential interruptions in auditing services if the TPA is under attack. Furthermore, users could exploit security issues to manipulate compensation from Cloud Service Providers (CSPs).

Global distributed file system construction has been the subject of much research, with mixed results. In the academic community, Andrew file system (AFS) is known as a very effective system. Over 100 million users can now be accommodated by industry platforms like BitTorrent, Napster, and KaZaA. The development of generic file systems that offer decentralized distribution, low latency, and global coverage is an ongoing endeavor [30]. From the client-server paradigm, which was originally developed for distributed systems with a smaller scale and a server with more processing power, the peer-to-peer (P2P) scheme evolved. Each node in a P2P network acts as both a client and a server, allowing for symmetric communication. Through the facilitation of direct communication between peers, P2P systems overcome bandwidth constraints in file sharing. The scalability and efficiency of file sharing are greatly improved when peers share files in parts instead of all clients requesting them from a server at once.

3.3 METHODOLOGY

The four primary components of the proposed system are shown in Figure 3.2: Data Owner Application (DOA), Data Consumer Applications (DCAs), Cloud Storage Service (CSS), and Blockchain. There are two subtypes of CSSs: private and public. Assuming a single DOA exists to generate data and upload it to the CSS, we will use this assumption throughout the study. A number of DOAs and DCAs are required for data integrity detection. Organizations that need the DIS can access it through the blockchain system, which requires them to launch a blockchain client on their own nodes. Any node can join or exit the blockchain network at any time.

Although the Cloud can also operate as a blockchain node in actuality, for the sake of simplicity in our proposed service structure, CSS specifically refers to CSS. Smart contracts hosted on the blockchain enable the DIS

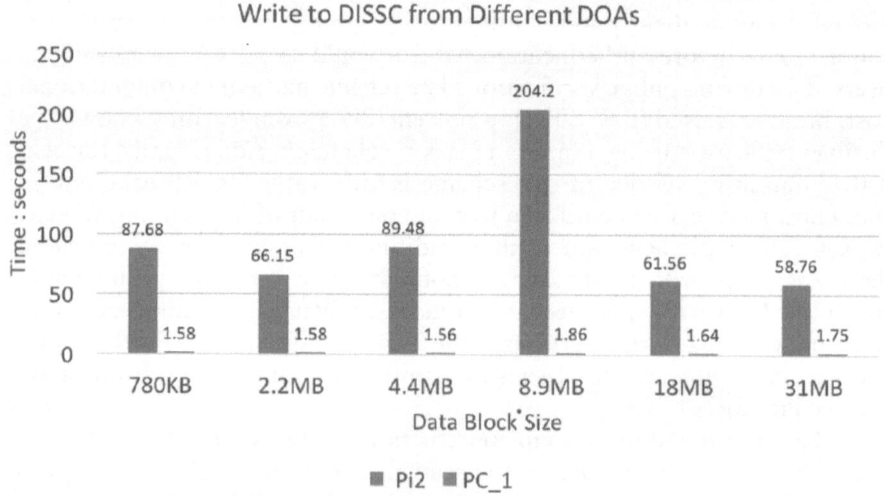

Figure 3.2 Contrasting the act of composing messages to DISSC originating from various Departure Operations Areas (DOA).

to be implemented. Improved DIS efficiency and dependability are results of this decentralized deployment. We make a number of assumptions to address concerns regarding the efficiency and security of the blockchain system, which have a major influence on our solution. The first presumption is that, since every node is in it for themselves, a 51% attack [31] is very unlikely to succeed. Consider the Bitcoin blockchain as an example; a hostile actor would be less likely to launch a 51% attack and more likely to engage in mining. The second premise is that reaching a consensus on a blockchain does not take long. Current blockchain consensus periods may be longer (16–18 seconds per block) or even fail under poor network conditions; however, Ethereum's blockchain is expected to enable consensus within an average length of 12.6 seconds. However, improvements in network and consensus algorithms may one day allow for the implementation of a consistently short time consensus.

3.3.1 Blockchain

Blockchain technology is the backbone of our proposed DIS, or Data Integrity Service. When the blockchain is first being set up, it is necessary for DOAs and DCAs to both join the network. This is the first step in creating a key pair, the public key of which will represent the node's blockchain account. In order to complete a transaction, the account must have enough gas, and the secret key is required to access the account. Assumption: The blockchain system can continue to function with a sufficient number of nodes ready to act as miners.

We implement a pay-per-transaction approach for the DIS by integrating blockchain into our platform. Only when the DOA needs to communicate with the smart contract does gas get used up. We found that this method greatly improved the DIS's flexibility when compared to our previous study, which used the cloud Information Management System (IMS) model [32]. Due to processing limits, DOAs may find it superfluous and tough to obtain gas by acting as miners. Thus, in our model, CSPs can continue to make money by acting as blockchain miners and earning gas, so there is no loss of profit for them. Soon, CSPs will be able to transact with DOAs using the gas they have earned. Then, DOAs can use their data to barter with DCAs for gas or incentives. DCAs can evaluate their hardware capabilities and financial situation to determine if they want to participate as miners.

3.3.2 Data Integrity Service (DIS)

A smart contract is used to implement the DIS. Before data can be securely stored in the blockchain by means of a smart contract, it must first be encrypted locally. All on-chain transactions involving the smart contract may be transparently audited because each party's account is used to engage with it. A node's blockchain data is synchronized with the entire blockchain network once the blockchain service (BS) is begun. When DOAs utilize the smart contract to insert data into the blockchain, the data does not become valid and available to other nodes until the blockchain reaches consensus. Read operations from the blockchain by DCAs via the smart contract, however, are lightning fast since DCAs are effectively reading data from their locally synced datasets. In comparison to the cloud-based IMS that was suggested in our earlier study, this feature makes our DIS more efficient [11]. DIS accessibility and flexibility are guaranteed with the deployment of the smart contract, which allows participants to engage with it at any moment. The only person or entity authorized to cancel or alter the DIS on the blockchain is the author. It is accessible at all times so long as the blockchain is running. The process of shutting down the blockchain system is made more complicated because, unlike services hosted by centralized providers, all participating nodes must stop providing BSs. Our proposed blockchain-based DIS outperforms the cloud-based IMS [11] in terms of efficiency and assurance.

Cloud Storage Service: Every major cloud computing service provider, including Amazon S3, IBM BlueMix, Microsoft Azure, and Digital Ocean, offers data storage services. These services are designed to cater to clients' economic capacities and application needs, providing flexible cloud storage solutions. In our framework, the CSS serves as a versatile data storage solution for DOs. Simultaneously, the Peer-to-Peer file system (P2PFS) facilitates data sharing between DOs and data consumers.

3.4 RESULTS AND DISCUSSION

The implementation of the proposed system is described in detail in this section. Our model is based on the proposed service framework, which is shown in Figure 3.3 for its complex structure. DOAs are compatible with both desktop computers and mobile devices connected to the Internet. Tasks such as data generation, uploading to CSS, hash generation, encryption of the resulting hash, and writing to the Data Integrity Service on the Blockchain (DISSC) are responsibilities of the devices' DOAs. At the same time, data integrity verification is done using DOAs on PCs. Desktop and cloud-based DCAs are required to be able to execute the P2PFS and the BS. BSs are definitely within the capabilities of modern PCs. We talk about single-board computers with General Purpose Inputs/Outputs (GPIOs) when we talk about IoT devices that gather data from sensors. Such features are included in many modern single-board computers. One example is the Raspberry Pi. Before sending data blocks to the CSS, IoT devices can process them directly from sensors. Ethereum is the most developed blockchain platform that currently supports smart contracts, hence it is used to create the blockchain system. One effort to transfer files in a manner similar to HyperText Transfer Protocol (HTTP) is IPFS (Inter-Planetary File System), which is used to implement the distributed

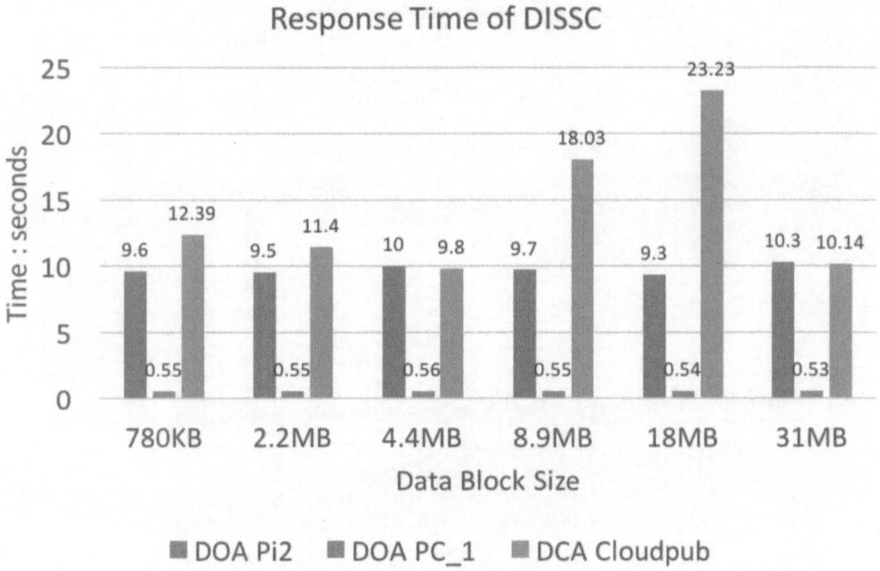

Figure 3.3 Time taken by DISSC to respond to both Data Owner Applications and Data Consumer Applications.

P2PFS. Protocol II and Protocol IV data verification is made easier with the use of IPFS.

In order to test the feasibility of our framework and related protocols, we have built a prototype system. Due to the ever-changing nature of IoT data, we assume that it is stored in data blocks of varying sizes, and that these blocks are combined to generate datasets. In most cases, you have to check the integrity of certain data blocks before you can guarantee the whole dataset is secure. Our current working system is built on a private blockchain that has at least four nodes. One of these nodes is in charge of hosting P2PFS, the blockchain, and CSS.

A personal computer is dual-tasked with running both DOA and DCA. Meanwhile, an IoT device is dedicated to running DOA for tasks such as data block generation and uploading to the CSS. However, a public cloud is assigned the role of running DCA, specifically to assess the efficiency of data block downloads under various network conditions. In Figure 3.1, the efficiency of verifying Data Integrity from DOA to CSS-Y is illustrated. The comparison results in Figure 3.2 indicate a substantially shorter time used by PC 1 compared to Pi2. This discrepancy can be attributed to at least two factors: differences in computation capability and variations in the Ethereum client version. Figures 3.3 and 3.4 depict the time spent by different DOAs and DCAs in querying the Data Integrity Service on the Blockchain (DISSC). Additionally, Figure 3.4

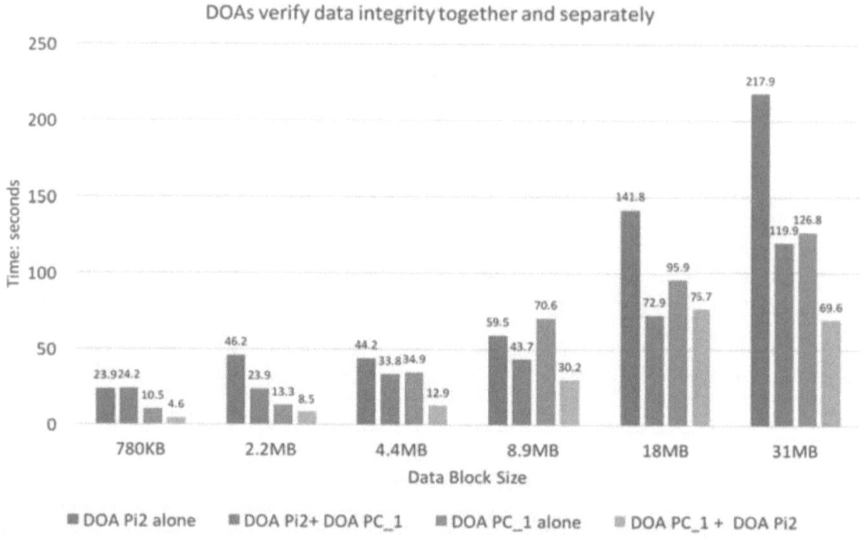

Figure 3.4 Duration required for the retrieval of data blocks.

outlines the time spent retrieving data blocks of different sizes in the CSS-N model. The test results demonstrate that our proposed framework can effectively support the integrity verification of data blocks by multiple DOAs or DCAs.

3.5 CONCLUSION

Cloud data security revolves around three fundamental aspects: confidentiality, integrity, and availability, commonly known as CIA. In our earlier research, we introduced the concept of Data Integrity as a Service to address concerns related to data integrity in CSSs. A notable drawback of existing techniques is the reliance on a trusted third-party authority for completing data integrity verification tasks. However, this assumption may not always hold true, leading to potentially untrustworthy results in data integrity verification. This paper introduces efforts to implement a blockchain-based DIS, providing several advantages over previous works. It enhances reliability, as the service cannot be terminated by a single cloud party. The efficiency of data integrity verification improves with an increasing number of clients. The proposed framework supports the trading of data with data consumers and implements a pay-per-transaction model for DIS.

REFERENCES

1. Jiao, Y., Wang, P., Niyato, D., & Suankaewmanee, K. (2019). Auction mechanisms in cloud/fog computing resource allocation for public blockchain networks. *IEEE Transactions on Parallel and Distributed Systems*, 30(9), 1975–1989.
2. Juels, A., & Kaliski Jr., B. S. (2007). PORs: Proofs of retrievability for large files. *Proceedings of the 14th ACM Conference on Computer and Communications Security*. ACM, Alexandria Virginia USA pp. 584–597.
3. Koblitz, N. (1987). Elliptic curve cryptosystems. *Mathematics of Computation*, 48(177), 203–209.
4. Ranjan, R., Pandey, D., Rai, A. K., Gupta, D., Singh, P., Kumar, P. R., & Mohanty, S. N. (2023). A manifold-level hybrid deep learning approach for sentiment classification using an autoregressive model. *Applied Sciences*, 13(5), 3091.
5. Li, J., Liu, Z., Chen, L., Chen, P., & Wu, J. (2017). Blockchain-based security architecture for distributed cloud storage. *2017 IEEE International Symposium on Parallel and Distributed Processing with Applications and 2017 IEEE International Conference on Ubiquitous Computing and Communications (ISPA/IUCC)*. IEEE, Guangzhou, China pp. 408–411.
6. Arivazhagan, N., Somasundaram, K., Mohammad, G. B., Kumar, P. R., et al. (2022). Cloud-Internet of Health Things (IOHT) task scheduling using hybrid moth flame optimization with deep neural network algorithm for e-healthcare systems. *Scientific Programming*, 2022, 1–12.
7. Li, K., Cheng, L., & Teng, C. I. (2020). Voluntary sharing and mandatory provision: Private information disclosure on social networking sites. *Information Processing & Management*, 57(1), 102128.
8. Li, X., Jiang, P., Chen, T., Luo, X., & Wen, Q. (2017). A survey on the security of blockchain systems. *Future Generation Computer Systems*, 107, 1–25.
9. Liang, X., Shetty, S. S., Tosh, D., Njilla, L., Kamhoua, C. A., & Kwiat, K. (2019). ProvChain: Blockchain-based cloud data provenance. *Blockchain for Distributed Systems Security*, 69, 468–477.
10. Lin, C., He, D., Huang, X., Choo, K. K. R., & Vasilakos, A. V. (2018). BSeIn: A blockchain-based secure mutual authentication with fine-grained access control system for industry 4.0. *Journal of Network and Computer Applications*, 116, 42–52.
11. Kumar, P. R., Mohammad, G. B., & Dileep, P. (2021). Real-time heart rate monitoring system using least square method. *Annals of the Romanian Society for Cell Biology*, 25(6), 16302–16308.
12. Merkle, R. C. (1987). A digital signature based on a conventional encryption function. *Conference on the Theory and Application of Cryptographic Techniques*. Springer, Santa Barbara, CA, USA pp. 369–378.
13. Miller, V. S. (1985). Use of elliptic curves in cryptography. *Conference on the Theory and Application of Cryptographic Techniques*. Springer, Santa Barbara, CA, USA pp. 417–426.

14. Muthanna, A., Ateya, A., Khakimov, A., Gudkova, I., Abuarqoub, A., Samouylov, K., et al. (2019). Secure and reliable IoT networks using fog computing with software-defined networking and blockchain. *Journal of Sensor and Actuator Networks*, 8(1), 15.
15. Mohammad, G. B., Selvarajan Shitharth, and Kumar, P. R. (2021). Integrated machine learning model for an URL phishing detection. *International Journal of Grid and Distributed Computing*, 14(1), 513–529.
16. Rass, S. (2013). Dynamic proofs of retrievability from chameleon-hashes. *2013 International Conference on Security and Cryptography (SECRYPT).* IEEE, Reykjavik, Iceland pp. 1–9.
17. Rawat, D. B., Doku, R., & Garuba, M. (2019). Cybersecurity in big data era: From securing big data to data-driven security. *IEEE Transactions on Services Computing*, 14(6), 2055–2072.
18. Kumar, P. R., & Ananthan, T. (2019). Machine vision using LabVIEW for label inspection. *Journal of Innovation in Computer Science and Engineering (JICSE)*, 9(1), 58–62.
19. Saggi, M. K., & Jain, S. (2018). A survey towards an integration of big data analytics to big insights for value-creation. *Information Processing & Management*, 54(5), 758–790.
20. Shacham, H., & Waters, B. (2008). Compact proofs of retrievability. *International Conference on the Theory and Application of Cryptology and Information Security*. Springer, Melbourne, VIC, Australia pp. 90–107.
21. Kumar, P. R. (2018). Wireless mobile charger using inductive coupling. *Journal of Emerging Technologies and Innovative Research (JETIR)*, 5(10), 40–44.
22. Ali, M. S., Vecchio, M., Pincheira, M., Dolui, K., Antonelli, F., & Rehmani, M. H. (2019). Applications of blockchains in the Internet of Things: A comprehensive survey. *IEEE Communications Surveys and Tutorials*, 21(2), 1676–1717, 2nd Quart.
23. Kumar, P. R. (2018). Position control of a stepper motor using LabVIEW. *3rd International Conference on Recent Trends in Electronics, Information Communication Technology (RTEICT),* Bangalore, India pp. 1551–1554.
24. Petersen, S., & Carlsen, S. (2011). WirelessHART versus ISA100.11a: The format war hits the factory floor. *IEEE Industrial Electronics Magazine*, 5(4), 23–34.
25. Mekki, K., Bajic, E., Chaxel, F., & Meyer, F. (2019). A comparative study of LPWAN technologies for large-scale IoT deployment. *ICT Express*, 5(1), 1–7.
26. Shilpa, B., Kumar, P. R., & Jha, R. K. (2023). Spreading factor optimization for interference mitigation in dense indoor LoRa networks. *IEEE IAS Global Conference on Emerging Technologies (GlobConET),* London, UK pp. 1–5.
27. Khutsoane, O., Isong, B., & Abu-Mahfouz, A. M. (2017). IoT devices and applications based on LoRa/LoRaWAN. In *IECON 2017 - 43rd Annual Conference of the IEEE Industrial Electronics Society,* Beijing, China, October/November, pp. 6107–6112.
28. Lu, X., Niyato, D., Jiang, H., Kim, D. I., Xiao, Y., & Han, Z. (2018). Ambient backscatter assisted wireless powered communications. *IEEE Wireless Communications*, 25(2), 170–177.

29. Zhou, J., Cao, Z., Dong, X., & Vasilakos, A. V. (2017). Security and privacy for cloud-based IoT: Challenges. *IEEE Communications Magazine*, *55*(1), 26–33.
30. Roman, R., Zhou, J., & Lopez, J. (2013). On the features and challenges of security and privacy in distributed Internet of Things. *Computer Networks*, 57(10), 2266–2279.
31. Kumar, P. R., & Shilpa, B. (2024). An IoT-based smart healthcare system with edge intelligence computing. In S. Satpathy, S. N. Mohanty, & S. Potluri (eds.), *Reconnoitering the Landscape of Edge Intelligence in Healthcare*. CRC Press: Boca Raton, FL, pp. 31–46.
32. Zheng, Z., Xie, S., Dai, H.-N., Chen, X., & Wang, H. (2018). Blockchain challenges and opportunities: A survey. *International Journal of Web and Grid Services*, 14(4), 352–375.

Chapter 4

Blockchain's impetus for secure IoT-enabled applications in smart city

Wasswa Shafik

4.1 INTRODUCTION

Smart cities symbolize a standard change in urban planning and administration, leveraging modern technology and information to enhance the wellness of their residents and foster lasting financial and social progression [1]. The intensifying process of urbanization in our culture, in which the majority of the international populace currently lives in city locations, has actually developed a pressing need for innovative techniques to deal with the complex concerns connected with city living [2]. Smart cities use an aggressive option to these troubles, advertising a unique stage of metropolitan progression noted by performance, link, and sustainability.

The essential concept underlying the smart city idea is integrating Information Communication Technologies (ICT) and the Web of Points (IoT) right into the real structure of the city framework utilizing the large information age. This combination helps with real-time information collection from a huge selection of resources, such as sensing units, electronic cameras, and various other IoT tools [3]. Because of this, it institutes an all-inclusive electronic semantic network, boosting decision-making, enhancing source allowance, and boosting the alternative metropolitan experience. Leveraging this large storage tank of info, communities hold the possibility to improve the quality and efficiency of civil services, covering different domain names like transport, medical care, power administration, and environmental durability [4].

The adaptation of these sophisticated modern technologies perfectly leads the way for building a comprehensive electronic structure that records instant information from varied beginnings, as provided in Figure 4.1, such as sensing units, cams, and a myriad of IoT gadgets [5]. Integrating Blockchain modern technology provides a durable device, protecting the sacredness and privacy of information, thus expertly securing it versus any immoral changes or violations. Consequently, this step-by-step harmony equips important understandings for educated, data-driven decision-making, cautious source appropriation, and improving the alternative metropolitan atmosphere [6].

DOI: 10.1201/9781032663005-4

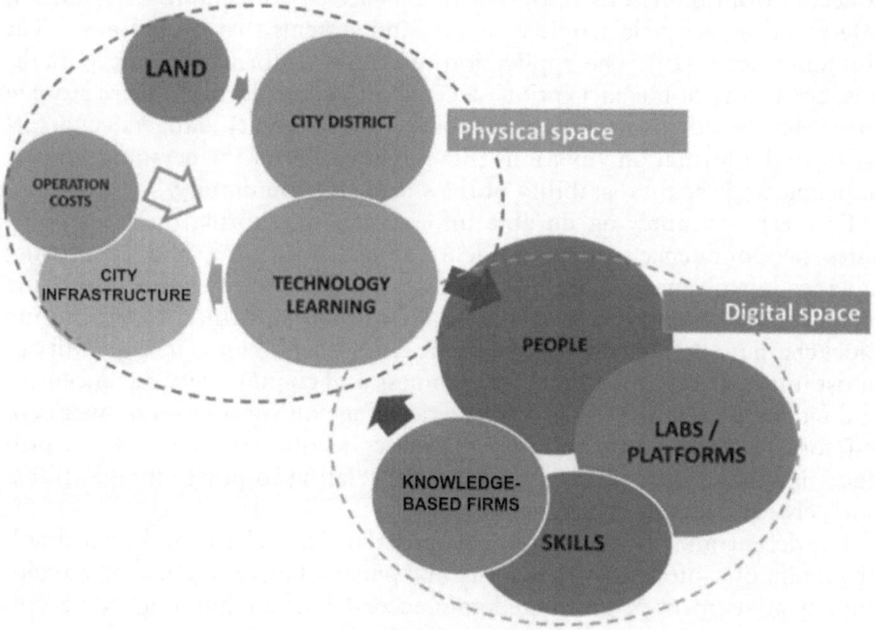

Figure 4.1 A cohesive range of bids for smart cities.

A varied range of innovations, from ecological sensing units to smart traffic signals, comprises IoT tools, jointly accumulating substantial information. This information, consequently, returns very useful understandings right into diverse facets of city presence. Guaranteeing this information's stability, exposure, and discretion is its protected storage space on a Blockchain system. Basic to Blockchain modern technology, wise agreements effortlessly promote automated, protected, and clear purchases and contracts [7]. Their applicability covers numerous domain names, consisting of power usage, waste administration, and public transportation solutions.

The wise city principle is based on incorporation and sustainability, focusing on reasonable and equivalent accessibility to innovation and solutions for all neighborhood participants. The ability of modern Blockchain technology to approve people's authority over their information and deals remains in conformity with the concepts of addition and personal privacy [8]. IoT-enabled gadgets have the prospective to proactively consist of people in decision-making procedures, allowing them to voluntarily add their information in return for boosted solutions, all while guarding their privacy [9].

The requirement for safe information administration is vital in the context of wise cities, where huge quantities of information are produced and made use of for metropolitan development and administration. Largely, it ensures the conservation of delicate individual and public information, thus promoting the personal privacy and safety of people. Information is collected from numerous resources in a smart city, including IoT gadgets, safety and security electronic cameras, and systems that promote resident communication [10]. The application of durable information safety methods, consisting of file encryption, accessibility controls, and secure storage space, is needed to minimize unapproved accessibility dangers, cyberattacks, and information violations that might endanger the personal privacy of people and the total stability of the smart city community.

Furthermore, applying durable information monitoring methods cultivates self-confidence among residents and stakeholders, and advertising enhanced involvement in data-sharing ventures is important for improving metropolitan solutions and facilities. The reasoning behind incorporating Blockchain modern technology in smart cities is based upon its capability to boost information protection and openness and count on city advancement and federal government [11]. Smart cities hinge on substantial networks of IoT tools, sensing units, and information collection terminals, which produce significant quantities of information relating to power intake, transport, civil services, and other elements.

The decentralized and unalterable journal of Blockchain modern technology maintains information's honesty and personal privacy, efficiently avoiding any adjustment or prohibited gain access. This method not only grows boosted guarantee among citizens and stakeholders concerning information accuracy but also enhances the application of data-driven decision-making treatments [12]. Using modern Blockchain technology allows the assistance of safe, clear, and auditable deals via smart agreements, therefore improving the performance of city procedures, advertising cost-efficient solutions, and promoting count on electronic communications [13]. Subsequently, this improvement adds to understanding the smart city principle, identified by lasting, effective, and comprehensive metropolitan living.

4.1.1 The chapter contribution

This study contributes the following as summarized:

- The study elucidates the profound impact of the IoT on the transformation of urban landscapes. It also scrutinizes the diverse range of applications of the IoT in smart cities, explicitly exploring the utilization of IoT in areas like traffic control, waste management, and energy efficiency.
- It provides an overview of Blockchain technology, including its principles and features, explaining the concept of decentralization, immutability, and consensus mechanisms in Blockchain.
- It discusses how Blockchain technology can enhance the security and reliability of IoT systems by exploring use cases of integrating Blockchain with IoT in smart cities, such as data integrity, device authentication, and secure transactions.

- It highlights the security and privacy challenges in IoT-enabled smart cities, enlightening on how Blockchain can address these concerns, including data encryption and access control.
- It demonstrates the security and privacy challenges in IoT-enabled smart cities, detailing how Blockchain can address these concerns, including data encryption and access control.
- It identifies the challenges and limitations of using Blockchain in smart cities, addressing scalability issues, energy consumption, and regulatory concerns.
- Finally, it ventures into the future of Blockchain and IoT in smart cities, discussing potential advancements and emerging trends.

4.1.2 The chapter organization

Section 4.2 presents the IoT in smart cities, explains the significance of IoT in transforming urban environments, and discusses the various applications of IoT in smart cities. Section 4.3 provides an overview of Blockchain technology, including its principles and features, and illustrates the concept of decentralization, immutability, and consensus mechanisms in Blockchain. Section 4.4 describes the security and privacy challenges in IoT-enabled smart cities and elucidates how Blockchain can address these concerns, including data encryption and access control. Section 4.5 identifies the challenges and limitations of using Blockchain in smart cities and addresses scalability issues, energy consumption, and regulatory concerns. Section 4.6 speculates on the future of Blockchain and IoT in smart cities and discusses potential advancements and emerging trends. Finally, Section 4.7 presents the conclusion.

4.2 INTERNET OF THINGS AND BLOCKCHAIN IN SMART CITIES

This section clarifies the relevance of IoT in remodeling lasting city and town settings with an easy conversation of the numerous applications of IoT in wise cities.

4.2.1 Internet of Things in smart cities

IoT matters in improving city settings by changing just how cities are constructed, carried out, and experienced. A number of critical variables show its impact, as listed below.

4.2.1.1 Data-driven decision-making

Making use of the IoT in information events and evaluations offers city authorities the ability to choose based on empirical proof, hence boosting the efficiency of urban planning and monitoring. By continuously checking and assessing information accumulated from different sensing units and gadgets, urban places quickly respond to vibrant scenarios, such as enhancing website traffic patterns to reduce blockage or reapportioning sources to locations needing prompt focus [13]. A data-driven technique makes it possible for metropolitan locations to react to vibrant difficulties quickly, therefore improving the general health of its residents.

4.2.1.2 Efficiency and sustainability

The IoT plays an important function in improving metropolitan sustainability with its capability to enhance source allotment efficiently. One instance of the possible advantages of wise waste monitoring systems is the capacity to enhance waste collection rounds by uniquely accumulating containers that have reached their complete ability [14]. This technique causes price financial savings and lowers gas use and carbon discharges. In a comparable capillary, a smart illumination system can regulate brightness levels adhering to the bordering light setting, therefore minimizing power usage. The performance enhancements are directly associated with sustainability, as they efficiently reduce a city's environmental influence and minimize the source problem, thus cultivating an ecologically mindful city setup [15].

4.2.1.3 Improved public services

The usage of the IoT in civil services causes a considerable improvement in how cities deal with the demands of their locals. The application of real-time information obtained from IoT sensing units allows the constant surveillance of top-quality water, thus assisting in the very early discovery of prospective worries and making certain safe alcohol consumption water [16]. Likewise, applying smart sensing units within fire hydrants autonomously informs pertinent authorities of possible emergencies, boosting the effectiveness of emergency action procedures. The improvements to civil services inevitably add to an enhanced lifestyle for metropolitan individuals as they experience the benefits of faster, more reputable, and more secure solutions [17]. Figure 4.2 shows estimations of IoTs and non-IoTs from 2015 to 2025.

4.2.1.4 Smart transportation

The IoT has actually produced an advanced makeover in the transport market, with its extensive results varying from maximizing web traffic circulation to boosting public transportation systems. Smart web traffic monitoring systems make use of information accumulated from a wide variety of sensing units to adjust traffic signals and reroute website

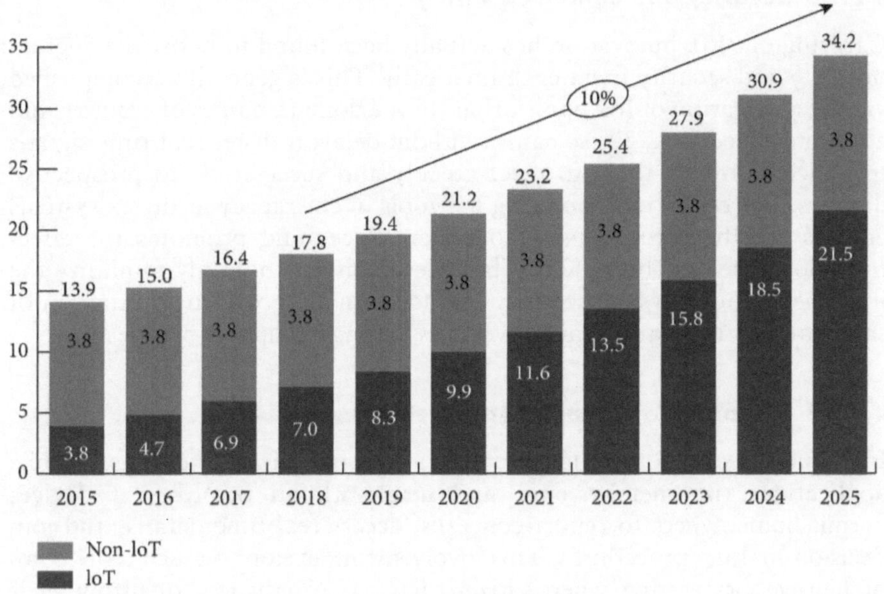

Figure 4.2 Total active device (non-Internet of Things and Internet of Things) connections worldwide [18].

traffic dynamically, therefore reducing blockage and lessening commute periods [17]. In addition, incorporating IoT innovation right into public transportation systems supplies travelers with prompt and current info, therefore boosting the effectiveness and ease of city traveling. Smart transport plays an important duty in affecting the future of metropolitan atmospheres by reducing traffic jams and motivating ecologically lasting transport choices [19].

4.2.1.5 Environmental monitoring

The IoT substantially resolves vital environmental monitoring worries, specifically in air and water top quality. Sensing units purposefully released around the city landscape constantly accumulate information concerning pollutants and environmental criteria. In circumstances where limits are gone beyond, governmental entities can, without delay, apply actions to reduce the damaging ecological repercussions and secure the wellness of the basic population [20]. The constant tracking of ecological problems offers the twin objective of protecting the natural environments and guarding the well-being of metropolitan citizens, thus coming to be an essential aspect in establishing a much healthier and much more lasting metropolitan atmosphere.

4.2.1.6 *Security and enhanced safety*

Carrying out IoT innovation has actually been found to boost the degrees of safety and security in cities dramatically. This is generally accomplished via the application of IoT applications in the domain names of security and emergency feedback. These cams, without delay, transfer real-time signals to police, allowing them to react quickly and successfully to prospective dangers. Moreover, incorporating IoT tools in emergency feedback systems enables specific recognition of occasion places and promotes the effective allowance of sources [21]. These developments not only minimize the improvement of physical security but additionally grow an assumption of safety and security and general wellness among occupants of cities.

4.2.1.7 *Sustainable citizen engagement*

IoT equips people to be energetic individuals in their city settings. Mobile applications, internet systems, and neighborhood involvement devices permit homeowners to report concerns, access real-time details, and join decision-making procedures. This involvement develops a much more comprehensive city setting where citizens have a straight risk of fitting their cities and enhancing the top quality of services, promoting a feeling of area participation and possession [22].

4.2.1.8 *Sustainable cost savings*

The long-lasting cost-saving possibility of the IoT offers a noteworthy benefit for urban areas. Executing effective source allowance techniques and the automation of procedures, such as anticipating upkeep for necessary framework, can produce substantial expense decreases in the future. Cities can successfully reduce the economic worry and functional hold-ups related to emergency repair services and solution disturbances by taking positive steps to recover facilities prior to getting to a state of failure [23]. In a comparable capillary, applying energy-efficient illumination and waste monitoring systems decreases functional expenditures, therefore helping with the extra efficient appropriation of sources by cities.

4.2.2 Blockchain technology in smart cities

Within this subsection, a summary of modern Blockchain technology, including its concepts and attributes, is supplied. It clarifies the principle of decentralization, immutability, and agreement devices in Blockchain. Using Blockchain innovation reinvents countless sectors with its capability to enhance openness, strengthen protection actions, maximize functional effectiveness, minimize expenditures, and grow trust funds [24].

The concepts and functions of Blockchain have substantial effects on the future of information monitoring and electronic trust funds, as seen by its noteworthy application in cryptocurrencies.

4.2.2.1 Core principles of Blockchain technology

Blockchain innovation works on a decentralized local area network, typically represented as nodes. Unlike traditional systems that rely upon a main authority or intermediary to validate and videotape deals, Blockchain innovation runs via a decentralized network of individuals.

4.2.2.1.1 Decentralization

The procedure of decentralization offers many benefits. Originally, making use of Blockchain innovation improves openness by making sure that all individuals within the network have equivalent accessibility to a common journal that is continually upgraded in real time. Moreover, it reinforces safety by removing any possible failing factors [24]. Rather than depending upon a particular entity, the confirmation of purchases is attained with the network's agreement, therefore considerably hampering the capability of a harmful star to regulate the system.

4.2.2.1.2 Distributed ledger

The basic concept underlying Blockchain innovation focuses on a decentralized journal. The journal acts as a detailed document of all purchases inside the network, arranged back to back via the development of blocks, where each block includes a collection of purchases. Linking blocks to develop a consecutive collection is in charge of the term "Blockchain." Every person within the network keeps the same reproduction of this journal [25]. The dispersed journal system ensures consistent accessibility to the deal background for all individuals, therefore dramatically hindering any efforts to change or remove information without getting agreement from the entire network.

4.2.2.1.3 Immutability

As soon as a deal is recorded in a block and added to the Blockchain, it ends up being exceptionally strenuous to modify or eliminate. The immutability of Blockchain innovation can be credited to the release of cryptographic hash features. Every block makes up a collection of deals and a special cryptographic hash of the coming before block. The procedure creates a safe series of interconnected blocks that, as soon as built, show a high level of resistance to any adjustment tries [26]. Customizing a solitary block requires modifying all adhering to blocks, a computationally not practical treatment.

4.2.3 Key features of Blockchain technology

4.2.3.1 *Security and transparency*

The high degree of safety and security in Blockchain is credited to its decentralized and unalterable qualities. Changing a solitary block would certainly demand the alteration of all succeeding blocks on the Blockchain, which is a computationally impractical job. Making use of modern Blockchain technology improves its durability versus illegal tasks and cyberpunk efforts. The integral openness of Blockchain innovation allows all individuals to gain access to and observe the total purchase background [27]. Although specific purchases are pseudonymous, implying alphanumeric addresses recognize them, the information continues to be available and proven. This attribute can be useful for objectives like bookkeeping and guaranteeing responsibility.

4.2.3.2 *Smart contracts and speed and efficiency*

The usage of modern Blockchain technology makes it possible to assist with wise agreements, which are agreements that are self-executing and have terms that are clearly inscribed right into the underlying code. These legal contracts are made to implement autonomously upon the gratification of pre-programmed situations, therefore reducing the need for intermediaries in varied functional treatments [28]. The application of modern Blockchain technology can substantially boost the speed and performance of deals, especially in cross-border repayments and supply chain administration.

4.2.3.3 *Accessibility*

The pledge of modern Blockchain technology expands past financing and causes revolutionary changes in a number of locations, such as health care, supply chain monitoring, electing systems, and others. The concepts of decentralization, immutability, and openness, paired with qualities such as wise agreements, add to the reinforcement of safety and security, effectiveness, and liability in varied applications [28]. Nonetheless, it is critical to challenge challenges such as scalability, power effectiveness, and governing structures to realize Blockchain modern technology's abilities totally.

4.3 INTEGRATION OF BLOCKCHAIN AND SECURE IOT APPLICATION

In the 21st century, wise cities have become a transformative force in contemporary urban planning. The intensifying international urbanization fad needs cutting-edge options to deal with the difficulties of traditional metropolitan facilities. Difficulties such as ineffective source monitoring, quick population growth, raised source needs, safety and personal privacy issues, and the importance of sustainability have actually pressed cities to discover sophisticated modern technologies [29]. Among these, Blockchain and IoT have actually appeared as appealing modern technologies, holding enormous possibilities to change smart city growth. Incorporating Blockchain and the IoT in smart cities stands for a standard change in urban planning and growth. We discover the combinations of these two transformative modern technologies in this area, diving right into their stamina and checking out just how their assimilation can deal with crucial obstacles traditional city facilities deal with. With a thorough evaluation of the advantages, this area intends to supply an understanding of the extensive influence of joining Blockchain and IoT together in wise cities [30].

4.3.1 Decentralized data management

Blockchain stands for an important structure for decentralized and tamper-proof information monitoring in smart cities. Incorporating it with IoT warranties that all deals and information originating from IoT gadgets are firmly tape-recorded on an unalterable journal, therefore guarding information honesty and preventing unapproved adjustments from harmful stars [31] Provided the extensive real-time information created by IoT sensing units and tools, Blockchain assists in safe administration without dependence on a main authority, mitigating information adjustment danger.

4.3.2 Secure identity and access management

Blockchain's safe and decentralized identification administration ability is a foundation for boosting protection and personal privacy within the wise city community (Khalil et al., 2022). An increased degree of safety is attained by enabling residents, gadgets, and entities to save their identifications on the Blockchain. This assimilation additionally addresses IoT gadgets' verification and permission demands, giving a durable structure for taking care of identifications and accessibility approvals [32]. This strategy makes certain safe and secure communications within the smart city network, strengthening the system's total durability.

4.3.3 Smart contracts for automation

Incorporating Blockchain and IoT presents smart agreements that automate and implement predefined guidelines, removing the demand for intermediaries. In wise cities, these agreements are essential in automating varied procedures such as energy settlements, waste administration, and traffic control. IoT tools with sensing units and actuators can cause occasions based on real-world information. As an example, integrating IoT and Blockchain-based agreements in tourist financing can improve procedures, instantly performing activities when details problems are fulfilled [33]. This assimilation improves performance and makes certain lasting economic techniques in the tourist sector.

4.3.4 Supply chain transparency

Blockchain's clear and deducible journal locates applications in supply chain monitoring, making sure the safe and secure recording of deals connected to manufacturing, transport, and circulation of items. The harmony with IoT sensing units in the supply chain allows real-time information generation on items' place, problem, and standing [34]. When combined with Blockchain, this information warrants openness and liability, alleviating the threat of scams and maximizing the total effectiveness of the supply chain.

4.3.5 Energy trading, data security and privacy

Blockchain militarizes peer-to-peer power trading, developing a decentralized and clear method for power purchases. In the IoT world, wise meters and sensing units keep track of power usage and manufacturing in structures. Incorporating this information with Blockchain makes it possible to make exact and clear payments and apply vibrant rate designs [35]. This urges power preservation and advertises making use of renewable resource resources, largely inside your home. The protected and decentralized Blockchain design boosts information safety and security, with cryptographic attributes making sure encrypted gain access just for licensed celebrations. The level of sensitivity of information accumulated by IoT gadgets makes protecting personal privacy extremely important [36]. Blockchain offers a decentralized control system, considerably lowering the threat of unapproved gain access to or meddling.

4.3.6 Citizen-centric services

Blockchain makes it possible for clear and protected resident solutions, extending tasks like a ballot, building enrollment, and health care documents. Blockchain's openness improves reliance on federal government solutions by equipping residents with better control over their information. Person interaction is additionally boosted with IoT tools accumulating comments and real-time details, with Blockchain combination ensuring the credibility of person input [37]. This cultivates even more answerable and receptive administration.

4.3.7 Real-time monitoring and analytics

As a safe and clear structure, Blockchain helps with the recording and sharing of real-time information. This comes to be specifically valuable in applications where information precision and stability are extremely important. IoT gadgets continually create real-time information on different city procedures. Combination with Blockchain allows the application of this information for analytics, decision-making, and optimization of city solutions, specifically in smart health care [38]. Therefore, cooperation between federal government companies, capitalism, and innovation companies is necessary to incorporate Blockchain and IoT in smart cities efficiently. Specifications and procedures for interoperability, safety, and information personal privacy should be developed to ensure a smooth and safe smart city ecological community.

4.4 IOT SECURITY AND PRIVACY CONCERNS

Incorporating Blockchain and IoT in wise cities ensures the transformation of information administration and administration. Nevertheless, this harmony presents a range of safety and security and personal privacy difficulties needing precise exams. This area divides these difficulties, using an academic expedition of the possible mistakes and needed safeguards.

4.4.1 Immutable data and access controls

The fundamental immutability of modern Blockchain technology, which ensures information honesty, comes to be a double-edged sword when taking into consideration the assimilation of Blockchain and IoT in smart cities. While this function safeguards versus information meddling, it increases considerable problems pertaining to the durability of delicate details. Within the complicated ecological community of a smart city, the lack of ability to customize or get rid of detailed information presents possible dangers to long-lasting personal privacy for people [38]. Creating durable gain access to controls and privacy-preserving devices is critical to browsing this difficulty successfully.

4.4.2 Smart contract vulnerabilities

Executing smart contracts, a foundational element of Blockchain-enabled IoT applications, introduces vulnerabilities that malicious actors may exploit. In smart cities, flaws in smart contracts can lead to unauthorized access, manipulation of data, or disruptions in automated processes, mirroring the concerns observed in web applications. To fortify smart contracts against potential threats, a comprehensive approach involving rigorous code audits and implementing secure coding practices is essential [39]. This proactive strategy is paramount in ensuring the integrity of the automated processes crucial for the seamless functioning of smart city infrastructures.

4.4.3 Scalability and network consensus

As the variety of IoT tools and purchases rises, scalability concerns and the choice of reliable agreement systems come to be the main issues. Ineffective agreement formulas and scalability traffic jams can endanger smart city facilities' safety and real-time responsiveness. Dealing with these obstacles calls for cutting-edge services that boost the scalability and performance of Blockchain networks [40]. By consistently developing the innovation underpinning these networks, we can much better suit the boosting needs of a growing wise city environment.

4.4.4 Network security and consensus vulnerabilities

The dependence on agreement formulas to verify purchases within Blockchain networks presents susceptibilities, specifically in large IoT implementations. Harmful stars making use of these susceptibilities can endanger the honesty of deals and keeping information. Advanced agreement systems and routine safety and security audits are crucial to strengthen network safety and security to neutralize this threat [41]. Aggressive steps are important to ensure the toughness of the agreement devices, guarding the wise city facilities against possible assaults on the Blockchain network.

4.4.5 Supply chain security for IoT devices

Making certain of the safety and security of the whole supply chain for IoT tools in smart cities is a complex difficulty. Meddling or endangering IoT gadgets at any phase of the supply chain can present susceptibilities right into the Blockchain network, weakening the total safety of smart city systems. Applying rigorous supply chain safety and security actions and gadget attestation procedures is vital to alleviate this danger thoroughly [41]. An aggressive and watchful method throughout all supply chain phases is important to preserving the honesty of the linked gadgets within the smart city landscape.

4.4.6 Identity management

Incorporating IoT with Blockchain requires durable identification monitoring systems, elevating issues regarding possibly revealing directly recognizable details. Poor identification defense steps might bring about the concession of a person's privacy. As a result, executing privacy-centric identification services, such as zero-knowledge evidence, is essential to reduce these dangers efficiently [42]. By including innovative personal privacy innovations, wise cities can strike a fragile equilibrium in between the benefits of identification administration and the conservation of specific personal privacy.

4.4.7 Data linkability

The clear nature of Blockchain presents the threat of information linkability, permitting entities to associate and evaluate transactional information. Relentless information linkability endangers resident privacy and personal privacy within the smart city landscape [43]. Constant advancements in cryptographic methods and privacy-enhancing modern technologies are vital to counter this threat. These developments are crucial in combating the dangers connected with information linkability, making sure that smart cities can harness the advantages of Blockchain while guarding residents' privacy. Blockchain and IoT are modern technologies in the developing landscape of smart cities.

4.5 CHALLENGES, LIMITATIONS, AND IMPLICATIONS

Within this section, the challenges and limitations of using Blockchain in smart cities are presented, as well as some implications.

4.5.1 Challenges

4.5.1.1 Scalability

On the Blockchain front, the problem hinges on effectively managing the large number of purchases happening within a smart city's vibrant and interconnected setting. As the city's facilities end up being progressively digitized, the stress on the Blockchain network can impede purchase handling rate and general system efficiency [44]. All at once, the spreading of IoT tools aggravates scalability worries. The expanding variety of these interconnected gadgets, varying from sensing units to wise devices, increases the intricacy of handling their communications flawlessly. They collaborate information exchange and interaction among a substantial selection of real-time tools, requiring durable scalability services to avoid traffic jams and ensure the smooth operation of smart city systems.

4.5.1.2 Interoperability

This is a vital difficulty in incorporating Blockchain and IoT within wise cities that focus on the smooth interaction between varied IoT gadgets and

Blockchain systems. The detailed internet of interconnected gadgets with distinct methods and requirements poses a considerable obstacle. IoT tools create substantial quantities of information that should be effectively and safely connected to the Blockchain for handling. Accomplishing a standard interaction procedure that can fit the diversification of IoT gadgets is important [45]. Without a typical language, the prospective advantages of the consolidated modern technologies might be jeopardized, preventing the smart city's capability to harness the full power of Blockchain and IoT harmonies.

4.5.1.3 Security concerns

Safety and security are vital problems when incorporating Blockchain and IoT in smart cities. In the world of Blockchain, the innovation naturally uses durable protection with its decentralized and cryptographic concepts. Nevertheless, susceptibilities can arise in wise agreements, the self-executing codes on Blockchain, or within the network's agreement formulas [46]. However, IoT tools present one more layer of safety and security difficulties. Tool susceptibilities, otherwise effectively resolved, can be made use of by harmful stars, endangering the whole system's stability. In addition, information safety and security problems in the interaction between IoT gadgets and the Blockchain network might be subject to delicate info. Safe and secure interaction procedures and durable verification devices will be vital to reducing these threats.

4.5.1.4 Energy consumption

The power usage related to Blockchain's Proof-of-Work (PoW) agreement formulas is a considerable problem, especially when considering their prospective application in smart cities. PoW includes complicated mathematical problems miners should resolve to confirm deals and develop brand-new blocks. This procedure needs significant computational power, resulting in high power usage. In the context of smart cities, where sustainability is an essential emphasis, the ecological effect of PoW ends up being a remarkable restriction [47]. The energy-intensive nature of PoW can add to raised carbon impacts and threaten the objective of producing effective and green city atmospheres. Resolving this obstacle requires discovering different agreement systems recognized for their reduced power demands.

4.5.2 Limitations

4.5.2.1 Cost

The expense effects of applying and keeping a Blockchain framework present a considerable difficulty for wise cities. The expenditures sustained in establishing the required equipment, software program, and network facilities, combined with recurring functional and upkeep expenses, can stress the funds of community authorities. Smart city tasks commonly call for significant financial investments, and embracing Blockchain innovation presents added monetary dedication [48]. Expenses relate to the first release and consist of costs for making certain protection, scalability, and conformity in time. Resolving these economic factors is vital for effectively incorporating Blockchain in smart city efforts, requiring mindful budgeting and critical preparation to stabilize technical development with monetary duty.

4.5.2.2 Regulatory uncertainty

This poses a substantial challenge to the prevalent execution of Blockchain and IoT in smart cities. The lack of clear and standard policies develops an ambiance of uncertainty, inhibiting public and economic sectors from completely accepting these transformative innovations [49]. The elaborate nature of Blockchain and IoT applications demands thorough, lawful structures to attend to information personal privacy, protection, and interoperability problems. Without distinct policies, stakeholders might think twice about buying or embracing these advancements, being afraid of lawful issues. As a result, it is important for policymakers to collaboratively create and develop clear regulative standards that cultivate advancement while dealing with the distinct difficulties offered by the combination of Blockchain and IoT in the context of wise cities [50].

4.5.2.3 Adoption barriers

The fostering of Blockchain and IoT, modern technologies in smart cities, encounters substantial obstacles, mainly rooted in resistance to alteration and the need for a thorough overhaul of existing systems. Cities and metropolitan frameworks usually operate reputable structures, and presenting turbulent modern technologies like Blockchain and IoT needs an essential change in the way of thinking and framework [51]. Stakeholders might be reluctant to welcome these modifications as a result of problems concerning the intricacy of the combination, possible disturbances throughout the change, and the prices entailed. Getting rid of these fostering obstacles demands critical preparation, clear interaction of the advantages, and steady execution techniques that reduce the regarded dangers connected with such transformative innovations [52].

4.5.3 Implications

4.5.3.1 Data integrity

Information honesty is vital in smart cities, where IoT gadgets constantly create substantial quantities of info (Andoni et al., 2019). The combination of Blockchain and modern technology plays a critical function in strengthening information honesty within this vibrant ecological community. Blockchain's integral attribute of immutability makes sure that as soon as information is taped on the Blockchain, it cannot be changed or damaged. In the context of IoT gadgets, this immutability attribute is specifically essential as it safeguards the credibility and dependability of the information gathered [53]. Each deal or item of info from an IoT gadget is safely tape-recorded in a block, and the decentralized and dispersed nature of the Blockchain makes certain that no solitary entity has control over the whole system, decreasing the danger of unapproved modifications. This development depends on the information produced by IoT tools. It supplies a durable structure for numerous applications, such as wise agreements and clear data-sharing systems in smart city facilities [54]. The guarantee of information stability with Blockchain assimilation not only boosts the integrity of info but also adds to the general protection and reliability of smart city systems.

4.5.3.2 Decentralization

This plays an essential duty in strengthening the durability of smart city systems by carrying out Blockchain innovation. By dispersing the control and storage space of information throughout a network of nodes instead of depending on a main authority, Blockchain alleviates the threat of a solitary factor of failing [55]. In the context of wise cities, this decentralized style makes certain that if one node or part stops working, the system in its entirety continues to be functional. This boosted toughness is specifically critical in city settings where the integrity of information and systems is vital for different solutions, such as transport, power monitoring, and public security [18].

4.5.3.3 Efficiency gains

Making use of Blockchain and IoT innovations in wise cities can substantially boost functional effectiveness. With its decentralized, safe, and secure nature, Blockchain ensures the stability and immutability of information accumulated from IoT gadgets. This is especially important in a smart city context where substantial quantities of information are produced from varied resources such as sensing units, cams, and various other linked gadgets [56]. One crucial element adding to performance gains is the clear and computerized implementation of wise agreements. These self-executing agreements, made possible by Blockchain, can automate different procedures in wise cities, varying from power administration and garbage disposal to traffic control. For example, smart agreements can help with automated settlements for power use based upon real-time information from IoT-enabled meters, getting rid of the requirement for intermediaries and improving the invoicing procedure. Furthermore, the decentralized nature of Blockchain eliminates the dependence on a solitary main authority, decreasing the danger of system failings or information violations [18]. This decentralization, paired with the protected and clear audit path supplied by Blockchain, boosts the total integrity of wise city procedures.

4.5.3.4 Transparent governance

Clear administration is a crucial facet of incorporating smart agreements on the Blockchain within the context of wise cities. Smart agreements, self-executing items of code with predefined policies, present openness and automation that can reinvent administration procedures. The immutability of Blockchain guarantees that the terms inscribed in these agreements are tamper-proof, cultivating trust funds among stakeholders [5,9]. This openness minimizes the danger of illegal tasks and permits people and authorities alike to trace and confirm every action of a procedure. Automated implementation of agreements better enhances administration, minimizing the requirement for intermediaries and reducing the possibility of human mistakes [59]. As smart cities progress, executing wise agreements comes to be a keystone in developing liable and reliable administration frameworks, establishing a criterion for a brand-new period of clear and automatic public administration.

4.6 FUTURE DIRECTIONS

The future instructions in incorporating Blockchain and IoT in smart cities hold tremendous capacity for transformative developments. Research study undertakings need to think about discovering hybrid Blockchain services that integrate the stamina of public and personal Blockchains, offering a nuanced strategy for scalability and personal privacy problems. Checking out the harmony between Blockchain, IoT, and side computers can alleviate latency problems and boost real-time handling abilities. Resolving safety and personal privacy difficulties stays critical, concentrating on establishing ingenious methods to secure the delicate information produced by IoT gadgets.

Future research studies ought to likewise explore energy-efficient agreement devices to minimize the ecological influence of Blockchain in smart city applications. Systematizing interoperability in between varied IoT gadgets and Blockchain systems is essential for smooth combination. Discovering the application of smart contracts for automating city administration procedures and boosting openness in supply chain administration stands for appealing methods. In addition, scientists ought to add to the growth of flexible regulative structures to fit the progressing landscape of Blockchain and IoT innovations in smart cities. Interest in user-centric layout, moral factors to consider, and the assimilation of expert systems will jointly form a lasting and comprehensive future for smart city applications.

4.7 CONCLUSION

Blockchain modern technology is leading an electronic change, transforming how we see and take care of trust funds and information in a linked globe. Decentralization, dispersed ledger innovation, immutability, and agreement procedures are its keystones, changing deal confirmation and information safety and security. Decentralization permits peer-to-peer networks to accept and tape purchases without a main authority, testing ordered frameworks. This step will certainly make it possible for individuals to rely on each other straight, removing the intermediaries. The dispersed journal preserves an integrated and unalterable deal background, advertising openness, and dissuading scams. Blockchain's unalterable journal and cryptographic protection supply unequaled information stability and cyber security. Openness and smart agreements boost responsibility and automation throughout industries, altering industrial purchases. Blockchain's effectiveness can accelerate treatments, minimize functional prices, and promote advancement in money, supply chain monitoring, health care, and administration. Blockchain is currently changing markets, yet scalability, power intake, and lawful structures remain concerns. It guarantees the development of trust funds, openness, and effectiveness in our vibrant electronic landscape. It is the structure of a decentralized, safe, and trust-based future that will certainly change our electronic age with long-lasting effects.

REFERENCES

1. Aggarwal, S., Chaudhary, R., Aujla, G. S., Kumar, N., Choo, K. K. R., & Zomaya, A. Y. (2019). Blockchain for smart communities: Applications, challenges, and opportunities. *Journal of Network and Computer Applications*, 144. https://doi.org/10.1016/j.jnca.2019.06.018.
2. Ranjan, R., Pandey, D., Rai, A. K., Gupta, D., Singh, P., Kumar, P. R., & Mohanty, S. N. (2023). A manifold-level hybrid deep learning approach for sentiment classification using an autoregressive model. *Applied Sciences*, 13(5), 3091. https://doi.org/10.3390/app13053091.
3. Ahmed, I., Zhang, Y., Jeon, G., Lin, W., Khosravi, M. R., & Qi, L. (2022). A Blockchain and artificial intelligence-enabled smart IoT framework for a sustainable city. *International Journal of Intelligent Systems,* 37(9). https://doi.org/10.1002/int.22852.

4. Alamri, B., Crowley, K., & Richardson, I. (2022). Blockchain-based identity management systems in health IoT: A systematic review. *IEEE Access,* 10, 59612–59629. https://doi.org/10.1109/ACCESS.2022.3180367.
5. Alqarni, M. A., Alkatheiri, M. S., Chauhdary, S. H., & Saleem, S. (2023). Use of Blockchain-based smart contracts in logistics and supply chains. *Electronics,* 12(6), 1340. https://doi.org/10.3390/electronics12061340.
6. Andoni, M., Robu, V., Flynn, D., Abram, S., Geach, D., Jenkins, D., McCallum, P., & Peacock, A. (2019). Blockchain technology in the energy sector: A systematic review of challenges and opportunities. *Renewable and Sustainable Energy Reviews,* 100. https://doi.org/10.1016/j.rser.2018.10.014.
7. Arivazhagan, N., Somasundaram, K., Mohammad, G. B., Kumar, P. R., et al. (2022). Cloud-Internet of Health Things (IOHT) task scheduling using hybrid moth flame optimization with deep neural network algorithm for e-healthcare systems. *Scientific Programming*, 2022, 1–12. https://doi.org/10.1155/2022/4100352.
8. Bandara, E., Shetty, S., Rahman, A., Mukkamala, R., & Liang, X. (2022). Moose: A scalable Blockchain architecture for 5G enabled IoT with sharding and network slicing. *IEEE Wireless Communications and Networking Conference, WCNC,* 2022, 1194–1199. https://doi.org/10.1109/WCNC51071.2022.9771885.
9. Bansod, S., & Ragha, L. (2022). Challenges in making Blockchain privacy compliant for the digital world: Some measures. *Sadhana - Academy Proceedings in Engineering Sciences,* 47(3), 1–17. https://doi.org/10.1007/S12046-022-01931-1.
10. Bao, Z., He, D., Wang, H., Luo, M., & Peng, C. (2023). A group signature scheme with selective linkability and traceability for Blockchain-based data sharing systems in e-health services. *IEEE Internet of Things Journal.* https://doi.org/10.1109/JIOT.2023.3284968.
11. Kumar, P. R., Mohammad, G. B., & Dileep, P. (2021). Real-time heart rate monitoring system using least square method. *Annals of the Romanian Society for Cell Biology*, 25(6), pp. 16302–16308. http://annalsofrscb.ro/index.php/journal/article/view/8878.
12. Bommu, S., Aravind Kumar, M., Babburu, K., Srikanth, N., Thalluri, L. N., Ganesh, V., Gopalan, A., Mallapati, P. K., Guha, K., Mohammad, H. R., & Kiran, S. (2023). Smart city IoT system network level routing analysis and Blockchain security based implementation. *Journal of Electrical Engineering and Technology,* 18(2). https://doi.org/10.1007/s42835-022-01239-4.
13. Centobelli, P., Cerchione, R., Vecchio, P. D., Oropallo, E., & Secundo, G. (2022). Blockchain technology for bridging trust, traceability and transparency in circular supply chain. *Information & Management,* 59(7), 103508. https://doi.org/10.1016/j.im.2021.103508.
14. Chakravarthi, P. K., Yuvaraj, D., & Venkataramanan, V. (2022). IoT-based smart energy meter for smart grids. *ICDCS 2022 - 2022 6th International Conference on Devices, Circuits and Systems,* pp. 360–363. https://doi.org/10.1109/ICDCS54290.2022.9780714.
15. Demirel, E. (2023). Application of Blockchain-based smart contract in sustainable tourism finance. *Blockchain for Tourism and Hospitality Industries,* 122–138. https://doi.org/10.4324/9781003351917.

16. Dieye, M., Valiorgue, P., Gelas, J. P., Diallo, E. H., Ghodous, P., Biennier, F., & Peyrol, E. (2023). A self-sovereign identity based on zero-knowledge proof and Blockchain. *IEEE Access,* 11, 49445–49455. https://doi.org/10.1109/ACCESS.2023.3268768.
17. Mohammad, G. B., Shitharth, S., & Kumar, P. R. (2021). Integrated machine learning model for an URL phishing detection. *International Journal of Grid and Distributed Computing*, 14(1), 513–529. http://sersc.org/journals/index.php/IJGDC/article/view/35886.
18. Yang, Z., Jianjun, L., Faqiri, H., Shafik, W., Talal Abdulrahman, A., Yusuf, M., & Sharawy, A. M. (2021). Green internet of things and big data application in smart cities development. *Complexity,* 2021, 1–15. https://doi.org/10.1155/2021/4922697.
19. Dong, J., Song, C., Liu, S., Yin, H., Zheng, H., & Li, Y. (2022). Decentralized peer-to-peer energy trading strategy in energy Blockchain environment: A game-theoretic approach. *Applied Energy,* 325, 119852. https://doi.org/10.1016/j.apenergy.2022.119852.
20. Hassanein, A. A., El-Tazi, N., & Mohy, N. N. (2022). Blockchain, smart contracts, and decentralized applications: An introduction. *Implementing and leveraging Blockchain Programming,* 97–114. https://doi.org/10.1007/978-981-16-3412-3_6.
21. Jha, S. K. (2023). Application of Blockchain technology in libraries and information centers services. *Library Hi Tech News.* https://doi.org/10.1108/LHTN-02-2023-0020.
22. Kashif, M., & Kalkan, K. (2024). EPIoT: Enhanced privacy preservation based Blockchain mechanism for internet-of-things. *Computer Networks,* 238, 110107. https://doi.org/10.1016/j.comnet.2023.110107.
23. Kumar, P. R., & Ananthan, T. (2019). Machine vision using LabVIEW for label inspection. *Journal of Innovation in Computer Science and Engineering (JICSE)*, 9(1), 58–62.
24. Khalil, U., Malik, O. A., Uddin, M., & Chen, C. L. (2022). A comparative analysis on Blockchain versus centralized authentication architectures for IoT-enabled smart devices in smart cities: A comprehensive review, recent advances, and future research directions. *Sensors,* 22(14), 5168. https://doi.org/10.3390/S22145168.
25. Khalil, U., Mueen-Uddin, Malik, O. A., & Hussain, S. (2022). A Blockchain footprint for authentication of IoT-enabled smart devices in smart cities: State-of-the-art advancements, challenges and future research directions. *IEEE Access,* 10. https://doi.org/10.1109/ACCESS.2022.3189998.
26. Kumar, P. R. (2018). Wireless mobile charger using inductive coupling. *Journal of Emerging Technologies and Innovative Research (JETIR)*, 5(10), 40–44.
27. Kushwaha, S. S., Joshi, S., Singh, D., Kaur, M., & Lee, H. N. (2022). Systematic review of security vulnerabilities in Ethereum Blockchain smart contract. *IEEE Access,* 10, 6605–6621. https://doi.org/10.1109/ACCESS.2021.3140091.
28. Liu, H., Han, S., & Zhu, Z. (2023). Blockchain technology toward smart construction: Review and future directions. *Journal of Construction Engineering and Management,* 149(3). https://doi.org/10.1061/jcemd4.coeng-11929.

29. Liu, Y., Wang, K., Qian, K., Du, M., & Guo, S. (2020). Tornado: Enabling Blockchain in heterogeneous internet of things through a space-structured approach. *IEEE Internet of Things Journal,* 7(2). https://doi.org/10.1109/JIOT.2019.2954128.
30. Ma, N., Waegel, A., Hakkarainen, M., Braham, W. W., Glass, L., & Aviv, D. (2023). Blockchain+IoT sensor network to measure, evaluate and incentivize personal environmental accounting and efficient energy use in indoor spaces. *Applied Energy,* 332, 120443. https://doi.org/10.1016/j.apenergy.2022.120443.
31. Shilpa, B., Kumar, P. R., & Jha, R. K. (2023). Spreading factor optimization for interference mitigation in dense indoor LoRa networks. *IEEE IAS Global Conference on Emerging Technologies (GlobConET),* London, UK, pp. 1–5.
32. Majeed, U., Khan, L. U., Yaqoob, I., Kazmi, S. M. A., Salah, K., & Hong, C. S. (2021). Blockchain for IoT-based smart cities: Recent advances, requirements, and future challenges. *Journal of Network and Computer Applications,* 181, 103007. https://doi.org/10.1016/j.jnca.2021.103007.
33. Makani, S., Pittala, R., Alsayed, E., Aloqaily, M., & Jararweh, Y. (2022). A survey of Blockchain applications in sustainable and smart cities. *Cluster Computing,* 25(6), 3915–3936. https://doi.org/10.1007/S10586-022-03625-Z.
34. Makhdoom, I., Abolhasan, M., Franklin, D., Lipman, J., Zimmermann, C., Piccardi, M., & Moghadam, N. S. (2023). Detecting compromised IoT devices: Existing techniques, challenges, and a way forward. *Computers & Security,* 132, 103384. https://doi.org/10.1016/j.cose.2023.103384.
35. Nica, E., Popescu, G. H., Poliak, M., Kliestik, T., & Sabie, O. M. (2023). Digital twin simulation tools, spatial cognition algorithms, and multi-sensor fusion technology in sustainable urban governance networks. *Mathematics,* 11(9). https://doi.org/10.3390/math11091981.
36. Nigmatov, A., Pradeep, A., & Musulmonova, N. (2023). Blockchain technology in improving transparency and efficiency in government operations. *15th International Conference on Electronics, Computers and Artificial Intelligence, ECAI 2023 - Proceedings.* https://doi.org/10.1109/ECAI58194.2023.10194154.
37. Kumar, P. R. (2018). Position control of a stepper motor using LabVIEW. *3rd International Conference on Recent Trends in Electronics, Information Communication Technology (RTEICT),* Bangalore, India, pp. 1551–1554.
38. Politou, E., Alepis, E., Virvou, M., & Patsakis, C. (2022). Privacy in Blockchain. In *Privacy and Data Protection Challenges in the Distributed Era.* Learning and Analytics in Intelligent Systems, vol. 26. Springer, Cham, pp. 133–149. https://doi.org/10.1007/978-3-030-85443-0_7.
39. Politou, E., Casino, F., Alepis, E., & Patsakis, C. (2021). Blockchain mutability: Challenges and proposed solutions. *IEEE Transactions on Emerging Topics in Computing,* 9(4), 1972–1986. https://doi.org/10.1109/TETC.2019.2949510.
40. Rejeb, A., Rejeb, K., Simske, S. J., & Keogh, J. G. (2021). Blockchain technology in the smart city: A bibliometric review. *Quality and Quantity.* https://doi.org/10.1007/s11135-021-01251-2.

41. Rejeb, A., Rejeb, K., Simske, S., Treiblmaier, H., & Zailani, S. (2022). The big picture on the internet of things and the smart city: A review of what we know and what we need to know. *Internet of Things*, 19. https://doi.org/10.1016/j.iot.2022.100565.
42. Kumar, P. R., & Shilpa, B. (2023). An IoT-based smart healthcare system with edge intelligence computing. In Satpathy, S., Mohanty, S. N., & Potluri, S. (eds.), *Reconnoitering the Landscape of Edge Intelligence in Healthcare*. CRC Press, Boca Raton, FL. https://doi.org/10.1007/978-981-16-6301-7_4.
43. Sabrina, F., Li, N., & Sohail, S. (2022). A Blockchain based secure IoT system using device identity management. *Sensors*, 22(19), 7535. https://doi.org/10.3390/S22197535.
44. Shafik, W., & Kalinaki, K. (2023). Smart city ecosystem: An exploration of requirements, architecture, applications, security, and emerging motivations. In Reddy, K. H. K., Roy, D. S., Mishra, T. K., & Hussain, M. W. (eds.), *Handbook of Research on Network-Enabled IoT Applications for Smart City Services*. IGI Global, Hershey, PA, pp. 75–98. https://www.igi-global.com/book/handbook-research-network-enabled-iot/324384.
45. Singh, S., Sanwar Hosen, A. S. M., & Yoon, B. (2021). Blockchain security attacks, challenges, and solutions for the future distributed IoT network. *IEEE Access*, 9, 13938–13959. https://doi.org/10.1109/ACCESS.2021.3051602.
46. Singh, S., Sharma, P. K., Yoon, B., Shojafar, M., Cho, G. H., & Ra, I. H. (2020). Convergence of Blockchain and artificial intelligence in IoT network for the sustainable smart city. *Sustainable Cities and Society*, 63. https://doi.org/10.1016/j.scs.2020.102364.
47. Su, X., Hu, Y., Liu, W., Jiang, Z., Qiu, C., Xiong, J., & Sun, J. (2023). A Blockchain-based smart contract model for secured energy trading management in smart microgrids. *Security and Privacy*, e341. https://doi.org/10.1002/SPY2.341.
48. Tang, Y., Yan, J., Chakraborty, C., & Sun, Y. (2023). Hedera: A permissionless and scalable hybrid Blockchain consensus algorithm in multi-access edge computing for IoT. *IEEE Internet of Things Journal*. https://doi.org/10.1109/JIOT.2023.3279108.
49. Testi, N., Marconi, R., & Pasher, E. (2023). Exploring the potential of Blockchain technology for citizen engagement in smart governance. *Open Research Europe*, 3, 183. https://doi.org/10.12688/OPENRESEUROPE.16153.1.
50. Shafik, W. (2024). Introduction to ChatGPT. In Obaid, A. J., Bhushan, B., Muthmainnah, S. P., & Rajest, S. S. (eds.), *Advanced Applications of Generative AI and Natural Language Processing Models*. IGI Global, Hershey, PA, pp. 1–25. https://www.igi-global.com/chapter/introduction-to-chatgpt/335830.
51. Tokkozhina, U., Mataloto, B. M., Martins, A. L., & Ferreira, J. C. (2023). Decentralizing online food delivery services: A Blockchain and IoT model for smart cities. *Mobile Networks and Applications*. https://doi.org/10.1007/s11036-023-02119-5.
52. Uchani Gutierrez, O. C., & Xu, G. (2023). Blockchain and smart contracts to secure property transactions in smart cities. *Applied Sciences*, 13(1), 66. https://doi.org/10.3390/APP13010066.

53. Ullah, Z., Naeem, M., Coronato, A., Ribino, P., & De Pietro, G. (2023). Blockchain Applications in Sustainable Smart Cities. *Sustainable Cities and Society,* 97, 104697. https://doi.org/10.1016/J.SCS.2023.104697.
54. Umran, S. M., Lu, S. F., Abduljabbar, Z. A., & Nyangaresi, V. O. (2023). Multi-chain Blockchain based secure data-sharing framework for industrial IoTs smart devices in petroleum industry. *Internet of Things*, 24, 100969. https://doi.org/10.1016/J.IOT.2023.100969.
55. Verma, S., & Sheel, A. (2022). Blockchain for government organizations: Past, present and future. *Journal of Global Operations and Strategic Sourcing,* 15(3), 406–430. https://doi.org/10.1108/JGOSS-08-2021-0063.
56. Yu, Z., Song, L., Jiang, L., & Khold Sharafi, O. (2022). Systematic literature review on the security challenges of Blockchain in IoT-based smart cities. *Kybernetes,* 51(1). https://doi.org/10.1108/K-07-2020-0449.

Chapter 5

A lightweight Enhanced KP-ABE system cyber-physical security and privacy advancement in healthcare networks

R. Nareshkumar, S. Umarani, Bharathi V, P. Sasikumar, and P. Uma Maheswari

5.1 INTRODUCTION

The advancement and widespread use of cyber-physical systems (CPSs) have enabled physical equipment to perform five key activities through the utilization of computers and networks. These functions include computation, communication, precise control, remote coordination, and autonomy [1]. Implementing a CPS will significantly enhance the competitiveness of key industrial sectors, including the mobile healthcare network (MHN). The structure of the MHN is seen in Figure 5.1. In the MHN [2], the mobile device is tasked with collecting data and uploading it to the healthcare cloud server. The patient and doctor possess a substantial amount of data that

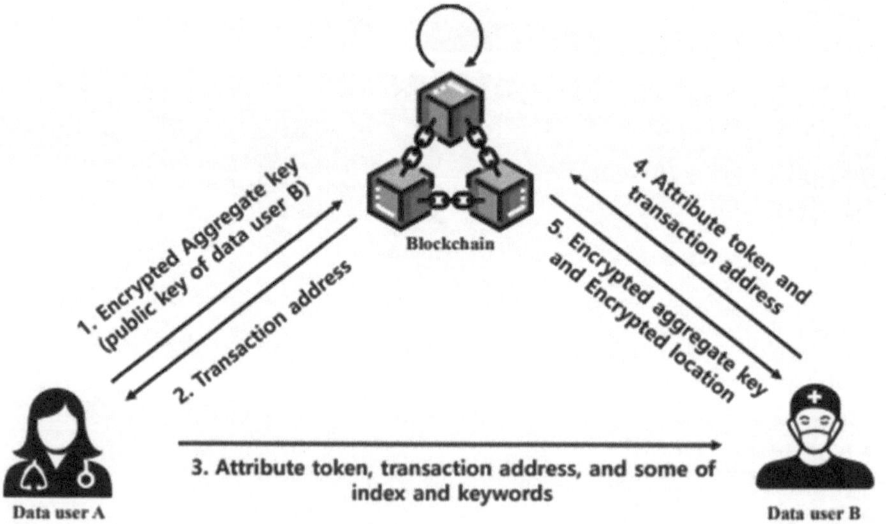

Figure 5.1 Data delegation [14].

DOI: 10.1201/9781032663005-5

needs to be stored and accessed on the healthcare cloud server. The hospital is responsible for overseeing the enrollment of mobile devices and operating the healthcare cloud server. The healthcare cloud server functions as the overseer of E-healthcare data, providing various services including data storage, uploading, and downloading. E-healthcare data in this system are a valuable resource for illness management, control, scientific research, and teaching. This has garnered increasing attention [3]. Data sharing has become an essential structural element in this setting, enabling real-time data interaction between medical professionals and patients, as well as providing real-time monitoring. The implementation of security measures for data sharing is crucial in order to prevent unauthorized user access and safeguard the data in the MHN. The ABE public key cryptosystem offers meticulous control over ciphertext access. The ABE system establishes a connection between the ciphertext and the key based on properties, allowing decryption only when a user's secret key aligns with the characteristics of the ciphertext. In the development of Key-Policy Attribute-Based Encryption (KP-ABE), a message is encrypted by using characteristics such as "profession: nurse, sex: female, and institution: hospital A," while keys are produced based on access regulations such as "profession: nurse $\wedge$ sex: female." Decryption of ciphertext is contingent upon the alignment of characteristics with the access policy. Ciphertext-policy attribute-based encryption (CP-ABE) is a variant of KP-ABE where the ciphertext is linked to an access policy and the key is linked to characteristics [4].

ABE is a valuable cryptographic technology that permits for the safe transfer of information to untrustworthy repositories, such as external web servers. ABE facilitates the efficient transmission of data provided among many stakeholders based on their respective roles or characteristics. Conventional encryption methods provide encryption from one end to another, but need individuals to supply decryption keys or store data in several encrypted copies with distinct keys. Neither choice is suitable. ABE may minimize the burden of key management in comparison to conventional encryption techniques. ABE offers precise access control and encryption that covers the whole communication process. Unscrupulous individuals may get access to encrypted data stored in the public repository that is not compatible with their confidential key. Nevertheless, in the absence of intelligible encryption keys, they are unable to comprehend the substance of the data. In addition, KP-ABE provides user privacy safeguards. In KP-ABE [5], a decryption user may access the authorized ciphertext even without knowledge of the encryptor. The data owner just has to identify the categories of users who are allowed to decode the ciphertext. Preserving the privacy and security of the MHN has evident advantages. When devices with limited resources are involved, there are certain problems that need resolution. In addition, devices with constrained resources include those with limited computational capabilities, storage capacity, and combined computing and storage capabilities, such as iPads and smartphones.

5.2 RELATED WORK HEALTHCARE AND ATTRIBUTE-BASED ENCRYPTION

An increasing number of programs are now prioritizing the issue of revocation. Created a reversible encryption system that enables indirect withdrawal, where each feature in the system has a specified time period of validity. The expert regularly informs the attributes and reallocates the user's key information.

5.2.1 Optimized variable attribute-based encryption

Y. Dong et al. [6] the study presents a prototype enhances the robustness of the blockchain scheme by protecting it from assaults on central points. The system employs an MA-ABE to obviate the need of a centralized authority, hence guaranteeing the security and confidentiality of the data stored on the blockchain. The authors assess the system's performance and showcase the practicality of constructing a redactable blockchain with access control. The viability of deploying a redactable blockchain with access control is shown, offering a pragmatic resolution for organizations seeking safe data exchange while retaining content control. The system's performance is assessed, demonstrating the practicality of implementing a redactable blockchain with access control.

The technology employs attribute-based encryption methods and a chameleon hash function to bolster security and privacy inside the blockchain system. The research results have significant ramifications for companies and sectors that handle confidential data, such as healthcare, banking, and government, where safeguarding privacy and ensuring data security are of utmost importance.

Roop Ranjan et al. [7], an innovative collaboration approach that text explores the implementation of blockchain as a means of exchanging messages and demonstrates that the access polynomial is an effective method for quickly revoking access in cloud storage. The research suggests using an access polynomial for key distribution, which minimizes the quantity of messages shared among group members. The suggested method offers a robust and distributed solution for sharing files in the cloud, guaranteeing the confidentiality of data and the anonymity of users. It allows for quick removal of user access without the need to communicate with users who have not had their access revoked, hence enhancing the efficiency of access control in cloud storage. The concept uses blockchain technology to enforce access control using a smart contract, hence removing the need for a trusted entity and mitigating denial of service threats. ABE allows the data owner to enforce access controls on files while maintaining user anonymity and improving privacy protection. The assessment findings indicate that the suggested system has a high level of scalability, making it appropriate for large-scale deployments involving up to 20,000 people. The incorporation of smart contracts in the suggested scheme eradicates any singular vulnerabilities and diminishes the expenses associated with system upkeep as compared to traditional alternatives.

L. Wu and J. Du's [8] research introduces an innovative access control strategy for implanted medical devices (IMDs) that relies on a proxy device, such as a smartphone, to do complex cryptographic calculations. This approach effectively extends the lifespan of the IMD. The proposed system uses to provide precise access control, guaranteeing that only qualified and authorized users may access the IMDs. The system employs CP-ABE to establish precise access control on the qualifications of the programming operator. The execution of the strategy on tangible emulation devices illustrates its viability and efficacy, offering a pragmatic resolution for safeguarding IMD. The technique is executed on actual emulator devices, and empirical findings show that it is efficient and impactful. The study showcases a functional model of the suggested approach by using genuine emulation devices, hence illustrating its practicality and efficiency.

M. Mahdavi et al.'s [9] research introduces KP-ABE techniques that use a reduced number of pairing processes in comparison to earlier KP-ABE schemes. Additionally, elliptic curve groups are utilized to guarantee shorter keys. Likewise, the study suggests safe techniques for delegating computationally intensive tasks, such as scalar multiplication by a curve point, exponentiation, and pairing, in fuzzy identity-based encryption (FIBE) and KP-ABE schemes. These solutions aim to enhance the ability of Internet of Things (IoT) devices to handle complicated operations. Furthermore, the article lacks a thorough assessment or examination of the suggested solutions in relation to their performance, efficiency, or feasibility in real-world IoT situations. Further study and testing may be required to determine the efficacy and practicality of the suggested methodologies in real IoT installations.

N. Arivazhagan et al. [10] introduce a new approach called CT-MA-ABE (Cross-Trust Multiple Authorization Attribute-Based Encryption) to tackle these problems. This method incorporates the function of a "notary" in cross-border exchanges, resolving the issue of oversight in completely decentralized alternatives while also taking into account the matter of confidence in centralized systems. The system's implementation in the legal jurisdictions of South China, including Zhuhai and Macau, serves as a crucial infrastructure component for ensuring the security of data exchanges. This successful deployment further highlights the system's efficacy as a dependable and secure solution.

F. Meng [11], an attempt to address this difficulty, academics have put up several Object-Oriented Attribute-Based Key Sharing (OOABKS) approaches. These approaches enhance the efficiency of creating ciphertexts online by pre-computing intermediate ciphertexts offline.

Nevertheless, we have identified several constraints in the current OOABKS schemes: (1) Certain schemes exhibit limited flexibility due to the inclusion of an access structure inside the intermediate ciphertext, rendering it incapable of generating a final ciphertext encrypted under a different access structure. (2) Alternative schemes are unable to meet the fundamental security prerequisite.

This study involves a thorough examination of current OOABKS systems. Here, we provide an improved OOABKS technique that provides verifiable security. Our technique preserves the benefits of online/offline ciphertext creation seen in prior OOAKBS systems while resolving their constraints. The experimental findings demonstrate that our modified system attains similar levels of efficiency as prior OOABKS schemes, while maintaining both flexibility and security.

PR. Kumar et al. [12] the current procedure for revoking most Revocable Attribute-Based Encryption (RABE) schemes is carried out by the cloud storage provider (CSP). As the CSP acts as an impartial and inquisitive intermediary, there is no assurance that the plaintext data associated with the revised ciphertext after revocation will remain unchanged. Furthermore, the majority of attribute-based encryption systems encounter problems associated with key escrow. In order to address the above-described problems, we propose a highly efficient RABE technique that not only ensures data integrity but also resolves the issue of key escrow.

PR. Kumar et al. [13] to be able to be effective, an Adaptive Secure Encryption (ASE) scheme must be capable of accommodating comprehensive search queries, which may be formulated as conjunctions, disjunctions, or any Boolean formulae. This study introduces a novel and efficient ASE algorithm, namely Feasible Approximate Solution Evaluation (FEASE), which offers both speed and expressiveness. The search process for every collection of keywords, regardless of its size, only takes three pairing operations. Additionally, the encryption and trapdoor methods have a linear complexity in relation to the amount of keywords. FEASE is built around a novel and efficient, which we provide as our first proposition. This scheme is noteworthy in its own right. In order to enhance security against keyword-guessing attacks, we expand the existing FEASE system to create the first and FEASE demonstrates superior performance compared to all currently available expressive ASE constructs.

J. Lee et al. [14] Personal Health Records (PHRs) store private information that can lead to privacy concerns. Moreover, in medical emergencies, it is important to consider multiple authorities to handle delegation. While different approaches are being researched for sharing data, they often fail to meet the required security standards in a real PHR sharing environment. This study introduces a system that utilizes key aggregate searchable encryption (KASE) to meet security requirements and utilizes blockchain and smart contracts to enhance data integrity, maintain data audit records, and ensure transparency. Researchers present a mechanism that guarantees the rights of individuals who possess PHR data when they delegate various rights using attribute tokens. We do both formal and informal security evaluations to assess the resilience of the proposed system against possible adversarial assaults.

PR. Kumar et al. [15] proxy re-encryption (PRE) enables a partially trusted intermediary, equipped with a re-encryption key, to convert a ciphertext encrypted using one key into an encryption of the same message using a different key. Attribute-based proxy re-encryption (AB-PRE) is an extension of PRE that allows for precise control over access to encrypted data and the ability to delegate access rights. Nevertheless, conventional AB-PRE is plagued by a single point of failure due to its dependence on a solitary proxy for executing ciphertext changes. In order to address this issue, this work presents a novel concept known as attribute-based threshold proxy re-encryption (AB-TPRE), which employs a number of proxies (N) for the purpose of transformations. In AB-TPRE, the re-encryption key is divided into N shares, with each proxy getting one share to produce a transformed ciphertext share. The converted ciphertext can only be successfully constructed when a certain number of transformed ciphertext shares, known as the threshold (t), are combined. In addition, we propose the implementation of a share updatable feature to mitigate the security risk associated with the potential leaking of shares. This property enables the refreshing of re-encryption key shares.

There is a notable absence of a comprehensive Systematic Literature Survey (SLR) in the current academic literature and research on the significance of ABE in health services and electronic health records (EHRs). Such a survey would provide valuable insights into the past, present, and ongoing advancements in ABE advances. It is well acknowledged among researchers that, before to starting any research project, an investigator consistently does a comprehensive literature review to choose the specific area of focus for future investigation. Therefore, it is crucial to conduct a comprehensive survey that encompasses all enhancements made to the ABE scheme up to the present day.

This literature review consolidates and outlines all ABE-based methodologies that have been specifically applied to health services since 2012. It includes updated versions and implementations across various specific categories, with the aim of employing explicit techniques to determine reliable conclusions based on these investigations. The main objective of this thorough and extensive study is to not only summarize existing research and literature, but also provide a framework for future research endeavors using ABE in the field of health services and other domains.

5.3 METHODOLOGY

The primary goal of an SLR is to comprehensively examine and include all relevant literature and studies related to a certain set of research topics and areas of interest. The research process often begins with a curiosity in a particular topic, but familiarity with the subject aids in formulating an appropriate research question for an inquiry. The investigators have utilized a two-stage screening method in order to extract the findings and papers from the databases. In the first stage, their bodies sort specific research based on Boolean-merged search strings, inclusion/exclusion criteria, and research concerns. In the second stage, they proceeded to extract the power source of the results and papers from the databases.

Within the context of this discussion, the word "Domain" refers to the region or domain in which a particular ABE system was presented and based on. When conducting this study, a total of seven different categories of domains are taken into consideration. These categories are as follows: "CP-ABE," "KP-ABE," "Hybrid," "Multiauthority-ABE," "Searchable Encryption-ABE," "Blockchain/Decentralized," and "Hierarchical ABE." As its name suggests, the "Hybrid" domain is a combination of two separate and different approaches that were not able to be accommodated inside the other domains (see Figure 5.1).

High-end granularity may be accomplished by the use of this method, which is known as Multiauthority-ABE. This approach allows for dispersed access rights to be granted to the data owner. Within the framework of the multi-authority structure, this part takes a look at a variety of different pieces of literature. Each study is emphasized and examined in accordance with the year in which it was published, beginning in 2013 and continuing until 2021. This is done while taking into consideration the key components of the various studies, such as the objectives, procedures, and enhancements that are proposed by the research.

Blockchain is a decentralized technology in Figure 5.2 that does not need a central authority to function. It was built on the foundation of distributed ledgers and consensus frameworks. As a result, a number of such alternatives that eliminate the need for attribute authority and analogous techniques are highlighted and discussed in the following paragraphs, beginning in the year 2019 and continuing through the year 2021.

5.3.1 A novel Key-Policy Attribute-Based Encryption (KP-ABE)

Here, researchers provide a novel KP-ABE [16] method that utilizes the concept of the Delegable identity-based broadcast encryption strategy to achieve ciphertexts of fixed size in Figure 5.3. The construction of KP-ABE is detailed as follows:

$$\mathrm{Setup}\left(1^{\lambda},\mathrm{U}\right)$$

The trusted attribute authority selects three cyclic groups, G_1, G_2, and G_T, of prime order p, with a bilinear pairing e: $G_1 \times G_2 \rightarrow G_T$, given the security parameter λ. The trusted attribute authority selects two generators g from G_1 and h from G_2, together with a secret value α from the set of non-zero integers modulo p, and a cryptographic hash function H that maps binary

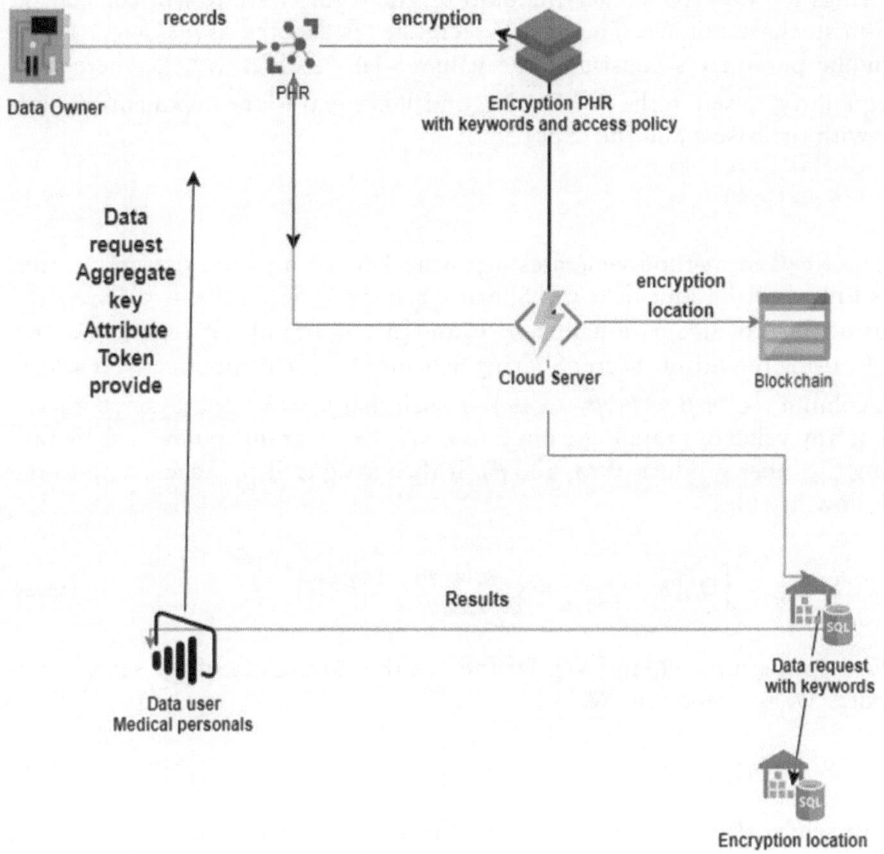

Figure 5.2 Proposed model architecture.

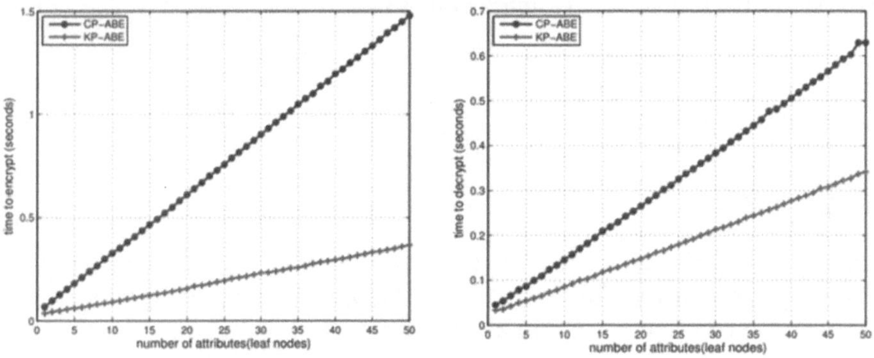

Figure 5.3 Data encryption and decryption.

strings to non-zero integers modulo p. The security study will consider H as a stochastic oracle. The master secret key is denoted as $msk=(g, \alpha)$. The public parameters consist of the values $=\left(w,v,h,h^{\alpha},\ldots,h^{\alpha^n}\right)$, where w is equal to g raised to the power of α, and V is equal to the exponentiation of e with the base g and the exponent h.

$$v=\hat{e}(g,h) \tag{5.1}$$

The KeyGen method generates a private key for an access structure that is linked to the Linear Secret Sharing Scheme (LSSS) scheme $(M_{\ell\times k},\rho)$ by using the provided parameters, msk, and (M,ρ), Initially, it produces shares of ℓ using the Linear Secret Sharing Scheme $(M_{\ell\times k},\rho)$. Specifically, it selects a column vector $\beta=(\beta_1,\beta_2,\ldots,\beta_k)^T$ T such that $\beta_1=s=1$ and $\beta_2,\ldots,\beta\in R\mathbb{Z}_p$. For any value of i ranging from 1 to $i=\ell$, the program computes λ_i by taking the inner product of M_i and β_T. It then assigns $\mathrm{SK}_{(M,\rho)}$ according to the following rules:

$$\mathrm{SK}_{(M,\rho)}=\left\{D_i,\left(K_{i,j}\right)_{j=1}^{n}\right\}_{i=1}^{\ell}=\left\{g^{\lambda_i/(\alpha+H(\rho(i)))},\left(h^{\lambda_i\alpha^j}\right)_{j=1}^{n}\right\}_{i=1}^{\ell}. \tag{5.2}$$

Encrypt(params, m, and **W**): Define t as the cardinality of the set of attributes **W**, and represent **W**.

$$\mathbf{W}=\{\omega_i\}_{i=1}^{t}. \tag{5.3}$$

The sender selects an element s from the set of non-zero integers RZ^*. Using this value, the sender calculates the ciphertext $c=(c_0, c_1, c_2)$ as follows: c_0 is obtained by multiplying m with the value V, c_1 is obtained by subtracting s from w, and c_2 is obtained by multiplying s with h and then multiplying the result with the product of t indices, where each index is obtained by adding α to the hash value of ω_i for each i from 1 to t.

$$c_0=m\cdot v^s=m\cdot\hat{e}(g,h)^s,$$
$$c_1=w^{-s}=g^{-\alpha s}, \tag{5.4}$$
$$c_2=h^{s\cdot\prod_{i=1}^{t}(\alpha+H(\omega_i))}.$$

Decryption of the ciphertext c, labeled with the set of characteristics $\mathbf{W}=\{\omega_i\}_{i=1}^{t}$, and using the secret key SK(M, ρ), involves parsing c as $c=(c_0$, c_1, and $c_2)$. The receiver initially defines the set I

$$\{i\rho(i)\in\mathbf{W}\} \tag{5.5}$$

and proceeds to compute the reconstruction constants

$$\{\mu_i\}_{i\in I} = \text{Recon}((M,\rho),W) \tag{5.6}$$

function. The decryption key for the LSSS scheme (M, ρ) is represented as

$$SK_{(M,\rho)} = \left\{D_i,(K_{i,j})_{j=1}^{n}\right\}_{i=1}^{\ell} \tag{5.7}$$

Next, the receiver does calculations.

$$p_{i,W}(\alpha) = \frac{\lambda_i}{\alpha}\left(\prod_{j=1,j\neq i}^{t}(\alpha + H(\omega_j)) - \prod_{j=1,j\neq i}^{t} H(\omega_j)\right) \tag{5.8}$$

$p_{i,W}(\alpha)$ is a polynomial of degree $t-2$ on the variable α, which is evident. The decrypting party can compute $h^{p_{i,W}(\alpha)}$ based on $(K_{i,j})_{j=1}^{n}$ for j ranging from 1 to n. Next, the decrypting party performs calculations.

$$Y_i = \left(\hat{e}\left(c_1, h^{p_{i,w}(\alpha)}\right)\cdot \hat{e}(D_i, c_2)\right)^{1/\prod_{j=1,j\neq i}^{t} H(\omega_j)} = \hat{e}(g,h)^{s\lambda_i} \tag{5.9}$$

Finally, the decrypting party computes.

$$Y = \prod_{i\in I} Y_i^{\mu_i} = \hat{e}(g,h)^s,\; m = \frac{c_0}{Y} \tag{5.10}$$

The suggested KP-ABE scheme is accurate.

Evidence. Let us assume that c is well-formed, indicating that c is encrypted using a certain set of properties.

$W = \{\omega_i\}_{i=1}^{t}$ then

$$\begin{aligned} Y_i &= \left(\hat{e}\left(c_1, h^{p_{i,w}(\alpha)}\right)\cdot \hat{e}(D_i, c_2)\right)^{1/\prod_{j=1,j\neq i}^{t} H(\omega_i)} \\ &= \left(\hat{e}(g,h)^{-s\alpha(\lambda_i/\alpha)\left(\prod_{j=1,j\neq i}^{t}(\alpha+H(\omega_j)) - \prod_{j=1,j\neq i}^{t} H(\omega_j)\right)} \hat{e}(g,h)^{s\lambda_i \prod_{i=1}^{t}(\alpha+H(\omega_j))}\right)^{1/\prod_{j=1,j\neq i}^{t} H(\omega_i)} \\ &= \left(\hat{e}(g,h)^{s\lambda_i \prod_{i=1}^{t} H(\omega_j)}\right)^{1/\prod_{j=1,j\neq i}^{t} H(\omega_i)} = \hat{e}(g,h)^{s\lambda_i}. \end{aligned} \tag{5.11}$$

Table 5.1 presents a comparison of the efficiency of several ABE schemes that are currently available, specifically focusing on methods that use

Table 5.1 Previous studies

Authors	*Methods*	*Contribution*	*Limitation*
Y. Dong et al. [6]	Decentralized consortium blockchain system	Prototype with redactability and access control using chameleon hash, digital signature, and multi-authority attribute-based encryption (MA-ABE). Enhances robustness and protects against central point attacks. Viability demonstrated in practical scenarios.	Limited assessment details on performance, efficiency, and real-world feasibility
A. Shafieinejad and Roop Ranjan et al. [7]	Blockchain and ABE for safe cloud file sharing	Innovative collaboration approach ensuring safe cloud file sharing. Uses access polynomial for quick access revocation. Incorporates smart contracts in blockchain for access control. High scalability.	Lack of detailed assessment on performance, efficiency, and real-world feasibility
L. Wu and J. Du [8]	Access control for implanted medical devices	Innovative access control strategy for implanted medical devices using proxy devices and ciphertext-policy attribute-based encryption (CP-ABE). Efficient execution on tangible emulator devices. Practical and efficient model demonstrated.	Limited information on scalability and real-world deployment
M. Mahdavi et al. [9]	Revised FIBE and KP-ABE techniques	Presents revised versions of fuzzy identity-based encryption (FIBE) and Key-Policy Attribute-Based Encryption (KP-ABE) schemes with reduced pairing processes. Lacks thorough assessment of performance in real-world IoT situations.	Lack of assessment details on performance, efficiency, and feasibility in real-world IoT situations
N. Arivazhagan et al. [10]	Cross-trust multiple authorization ABE	Introduces CT-MA-ABE for cross-border institutional authorization using MA-ABE and blockchain certification authority (BCA). Implements encryption-based authorization for privacy. Successful deployment in South China legal jurisdictions.	Successful deployment in specific regions, lack of broader assessment details
F. Meng [11]	Improved OOABKS with verifiable security	Thorough examination of current Object-Oriented Attribute-Based Key Sharing (OOABKS) systems. Presents an improved OOABKS technique with verifiable security and efficiency. Maintains flexibility and security.	Lack of information on specific constraints in current OOABKS schemes

(*Continued*)

Table 5.1 (Continued) Previous studies

Authors	*Methods*	*Contribution*	*Limitation*
PR. Kumar et al. [12]	Revocable Attribute-Based Encryption (RABE)	Proposes an efficient RABE technique addressing data integrity and key escrow issues. Security demonstrated using decisional q-parallel bilinear Diffie-Hellman exponent (q-PBDHE) assumption and discrete logarithm (DL) premise.	Lacks specific information on performance and scalability
PR. Kumar et al. [13]	Adaptive Secure Encryption (ASE)	Introduces FEASE, an Adaptive Secure Encryption (ASE) algorithm, providing speed and expressiveness. Enhances security against keyword-guessing attacks. Superior performance compared to existing ASE constructs.	Lacks detailed information on specific limitations and potential vulnerabilities
J. Lee et al. [14]	Key Aggregate Searchable Encryption (KASE)	Introduces a system for Personal Health Records (PHRs) that uses key aggregate searchable encryption (KASE) for security, and blockchain with smart contracts for data integrity, audit records, and transparency. Ensures rights of individuals during delegation using attribute tokens. Formal and informal security evaluations conducted.	Lacks detailed information on specific security evaluations, and potential challenges in real-world implementation
PR. Kumar et al. [15]	Attribute-Based Threshold Proxy Re-Encryption (AB-TPRE)	Presents attribute-based threshold proxy re-encryption (AB-TPRE) as a solution to the single point of failure in conventional Attribute-Based Proxy Re-Encryption (AB-PRE). AB-TPRE employs multiple proxies with a threshold mechanism for secure transformations. Introduces a share updatable feature to mitigate security risks.	The effectiveness and practicality of the proposed share updatable feature need to be thoroughly tested. Additional assessment details, especially in real-world scenarios, are required.

ciphertexts of a fixed size. Introduced a strategy for CP-ABE and KP-ABE that utilizes ciphertexts of constant size. We refer to these as schemes 1 and 2, respectively.

5.3.2 Security of BLS and Boneh-Gentry-Lynn-Shacham (BGLS) signature

An aggregate signature system is a cryptographic method that allows for the combination of several signatures on multiple messages from multiple users into a single concise signature.

Regrettably, there are currently no aggregate signature methods that are completely non-interactive, even when relying on the random oracle heuristic. This means that signers are need to exchange messages with each other, either in a sequential or alternative manner, in order to create the signature. Interacting with some intriguing apps might be too expensive.

Aggregate signatures, as defined by S. Ilakiya et al. [17], are a kind of digital signature that allows any party to consolidate many signatures on multiple messages from multiple users into a single concise signature. This basic element is valuable in several scenarios when there is limited storage or bandwidth capacity, and thus, there is a need to minimize the overall cryptographic burden.

Basic BGLS aggregate signature scheme [17]:

Set up	$G_1 = \langle g_1 \rangle, G_2 = \langle g_2 \rangle, G_T$, prime order p
	$\mathbf{H} : \{0,1\}^* \to G_1$, full domain
	$e : G_1 \times G_2 \to G_T$, bilinear
KeyGen	$x \leftarrow_{\$} \mathbb{Z}_p$
	$sk = x \in \mathbb{Z}_p$
	$pk = (y_1, y_2) = (g_1^x, g_2^x) \in G_1 \times G_2$
Sign (sk, m)	$\sigma = \mathbf{H}(m)^{sk} \in G_1$
$\text{Agg}(\sigma_1, \ldots, \sigma_k)$	$\sigma_A = \prod_{i=1}^{k} \sigma_i \in G_1$
$\text{AggVer}(\sigma_A, (pk_1, m_1), \ldots, (pk_k, m_k))$	$\prod_{i=1}^{k} e(\mathbf{H}(m_i), y_{2,i}) \stackrel{?}{=} e(\sigma_A, g_2)$

5.4 ENCRYPTION AND HEALTHCARE

The first factor that guarantees the unity and strength of a healthcare system is the protection of personal health information (PHI). The introduction of digital technology in healthcare allows for convenient access to this data, which facilitates the provision of more efficient and impactful health services and therapy. Presently, healthcare professionals widely use EHR technology. Most medical offices have shifted their EHR systems [18] to use outsourcing options, wherein EHR data is kept in centralized data warehouses on a third-party cloud that may lack total reliability and security.

Healthcare organizations must possess the capability to monitor and record user entry into their network. In order to do this, it is important to provide each user a distinct user ID for the purpose of accessing your organization's network. Ensuring the security of healthcare networks is a very successful strategy for preventing breaches. Malicious actors that successfully breach an organization's internal network have the capability to disseminate malware over the whole of the organization's system, therefore infecting every device that establishes a connection to the network. Failure to adopt efficient network management may have a negative impact on a firm.

Data backup and disaster recovery refer to the systematic practice of creating duplicate copies of crucial information and establishing a well-defined strategy for retrieving and restoring these files in the event of a catastrophe. HIPAA mandates the replication and storage of ePHI at an external facility. Healthcare organizations must also establish a disaster recovery strategy and provide a means of accessing data during emergency scenarios. In order to adhere to HIPAA regulations, it is essential to regularly create backups of data to mitigate the risk of patient data loss.

5.5 RESULTS AND DISCUSSION

The CP-ABE encryption technique, which is seen in Figure 5.4, is required to choose one random element from $\mathbb{Z}_p$ (during the building of polynomial p) and run two exponentiations in G (an elliptic curve group) for each attribute in order to encrypt the data. The KP-ABE encryption technique [19], however, only has to do two exponentiations in G (an elliptic curve group) for each attribute, and it only needs to choose one random element from $\mathbb{Z}_p$ across all of the attributes. Consequently, the time required for encryption by KP-ABE [20] is about three times quicker than the time required by CP-ABE [21].

In terms of decryption, the costliest calculation is the bilinear mapping, which is performed twice in CP-ABE and just once in KP-ABE for each leaf node. Consequently, the decryption time of KP-ABE is about two times quicker than the decryption time of CP-ABE [22]. However, while being more computationally efficient than CP-ABE in all three dimensions, KP-ABE has two notable drawbacks. Initially, it is important to note that the dimensions of the public key and master key increase proportionally with the quantity of characteristics in the overall set. This might result in an increase in communication overhead for the system, particularly when the system often necessitates users to update their public key. The relationship

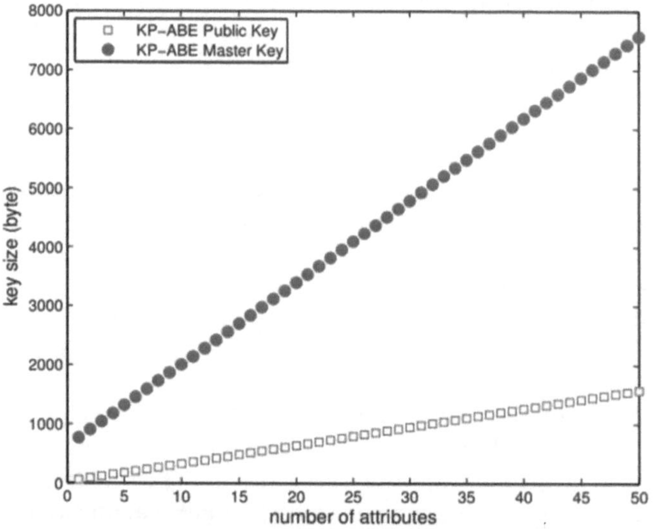

Figure 5.4 Enhanced Key-Policy Attribute-Based Encryption public and master key size.

between the sizes of a public key, a master key show in Figures 5.5 and 5.6, and qualities in the universe. Our method observes that a numerical attribute is represented by a maximum of 64 unique characteristics. The current KP-ABE [23] characteristics universe of a PHR system may be rather extensive. However, we may enhance the limitation on key size by using a big universe architecture of KP-ABE.

Comparisons are conducted based on the dimensions of the private key, the size of the ciphertext, and the quantity of pairing assessments during encryption and decryption given in Table 5.2.

An intriguing unresolved issue is the development of a KP-ABE (see Figures 5.4 and 5.5) [24] method that produces ciphertexts of a fixed size while maintaining security under a widely accepted premise or attaining a more robust definition of complete security. A further complex issue is the construction of a KP-ABE scheme that maintains a consistent size for both the ciphertext and the private key. We provide a synchronized aggregate signature construction in the random oracle model that is more efficient than our conventional model construction. Additionally, it has features that may make it more appealing for certain applications compared to current random oracle schemes. Synchronized aggregate signatures [17,25] may decrease the bandwidth demands that message signing places on a network. Intermediate routing nodes have the ability to aggregate signatures at any location when numerous signatures need to be sent to the collector, instead of carrying all signature data.

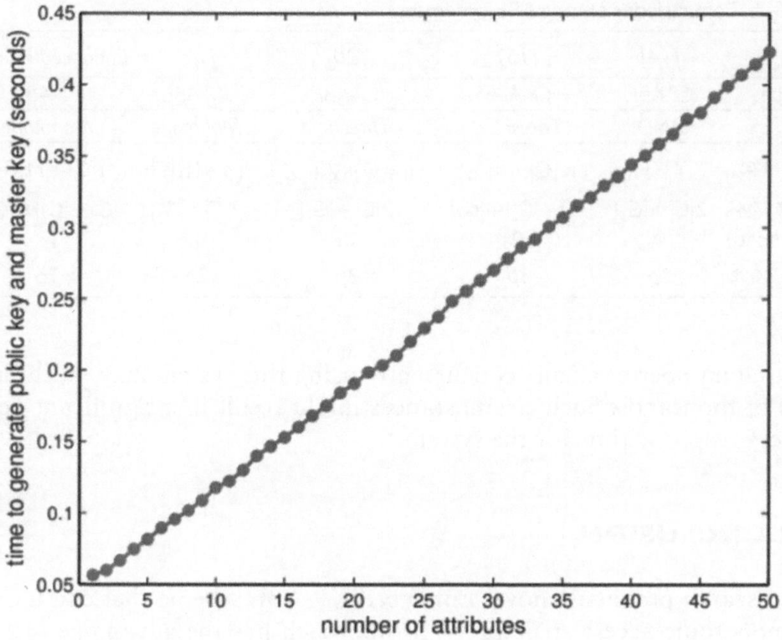

Figure 5.5 Enhanced Key-Policy Attribute-Based Encryption key size.

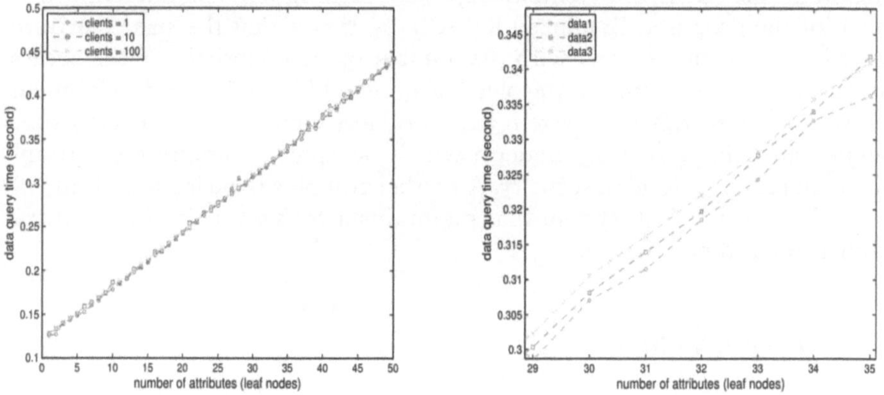

Figure 5.6 Data query number of classes.

The data retrieval time when numerous data consumers are involved is particularly intriguing. Figure 5.6 demonstrates that the time it takes to query data in a situation with several clients is directly proportional to the number of characteristics. However, this is not the case in a single client situation. The number of customers has a direct impact on the total data query time. Envisioning a nationwide deployment of a PHR system, the

Table 5.2 Comparisons among ABE schemes

	[14]	*[15]*	*[20]*	*[19]*	*Enhanced KP-ABE*
ABE Type	*CP-ABE*	*CP-ABE*	*CP-ABE*	*KP-ABE*	*KP-ABE*
Access	*AND*	*Threshold*	*Threshold*	*Monotone*	*Monotone*
Private key	$2\|G$	$w\|G_1\|+n\|G_2\|$	$(n+w)\|G\|+\|Z_p^*\|$	$(n+1)t\|G\|$	$t\|G_1\|+nt\|G_2\|$
Cipher text	$2\|G\|+\|G_T\|$	$\|G_1\|+\|G_2\|+\|G_T\|$	$2\|G\|+\|G_T\|$	$2\|G\|+\|G_T\|$	$\|G_1\|+\|G_2\|+\|G_T\|$
Encry cost	0	0	$1p$	$1p$	$1p$
Decrp cost	$2p$	$3p$	$3p$	$2p$	$2p$

potential number of clients concurrently using this system may reach hundreds of thousands. Such circumstances might result in a significant delay in query response time for the system.

5.6 CONCLUSION

This research presents a novel Enhanced KP-ABE scheme that can handle any monotonic access structure. The approach has the advantage of having ciphertexts of a fixed size. Additionally, we have demonstrated that the proposed scheme achieves semantic security in the selective-set model, relying on the generic Diffie-Hellman exponent assumption. An inherent drawback of the proposed Enhanced KP-ABE system is that the size of private keys increases significantly with the number of characteristics in the access structure. An interesting unresolved issue would include the development of an Enhanced KP-ABE system that produces ciphertexts of a fixed size, while maintaining security under a widely accepted assumption or attaining a more robust level of security. A further complex issue is the creation of an Enhanced KP-ABE system that maintains a consistent size for both the ciphertext and private key.

5.7 FUTURE WORK

Strengthen the ABE library by addressing the number of constraints and deficiencies in its collection including limited support for complex access structures inefficiencies in key generation and encryption processes inadequate handling of large attribute sets and a lack of optimization for scalability and performance in real-world applications. Our ABE library suffers from a number of constraints and deficiencies in its collection. The security advantage is considerably diminished by the amount of time that is required to use the ABE library. Because of this, we are going to make an effort to include new ABE architecture into our ABE library in order to improve its overall performance. Investigate ways to demonstrate access hierarchies that are both more animated and more effective. At the moment, we define access structures in a manner that is static, which is not appropriate for a dynamic system such as PHR. Additionally, a more effective way of description has the potential to minimize the complexity of both key management and policy. The combination of cryptographic techniques with additional methods that enhance privacy is recommended. A crucial way for protecting private data from cloud servers that are only partly trustworthy is the use of cryptographic algorithms; however, this is not the only option available. In light of this, we are working to discover a more effective method of addressing the problem of privacy and security in PHR systems.

REFERENCES

1. K. Sravanthi and P. Chandrasekhar, "An efficient multi-user groupwise integrity CP-ABE(GI-CPABE) for homogeneous and heterogeneous cloud blockchain transactions," *Journal of Electrical Systems*, vol. 20, no. 1, pp. 326–349, 2024, doi: 10.52783/jes.685.
2. S. Chawla and N. Gupta, "A proxy-based and collusion resistant multi-authority revocable CPABE framework with efficient user and attribute-level revocation (PCMR-CPABE)," *International Journal of Safety and Security Engineering*, vol. 13, no. 3, pp. 527–538, 2023, doi: 10.18280/ijsse.130315.
3. B. Shilpa, P. R. Kumar, and R. K. Jha, "LoRa DL: A deep learning model for enhancing the data transmission over LoRa using autoencoder," *The Journal of Supercomputing*, vol. 79, pp. 17079–17097, 2023.
4. H. Cui, Z. Wan, T. Zhaolu, H. Wang, and A. Miyaji, "Pay-per-proof: Decentralized outsourced multi-user PoR for cloud storage payment using blockchain," *IEEE Transactions on Cloud Computing*, pp. 1–14, 2023, doi: 10.1109/tcc.2023.3343710.
5. Y. Zhang, H. Geng, L. Su, and L. Lu, "A blockchain-based efficient data integrity verification scheme in multi-cloud storage," *IEEE Access*, vol. 10, pp. 105920–105929, 2022, doi: 10.1109/access.2022.3211391.
6. Y. Dong, Y. Li, Y. Cheng, and D. Yu, "Redactable consortium blockchain with access control: Leveraging chameleon hash and multi-authority attribute-based encryption," *High-Confidence Computing*, vol. 4, no. 1, p. 100168, 2024, doi: 10.1016/j.hcc.2023.100168.
7. R. Ranjan, D. Pandey, A. K. Rai, D. Gupta, P. Singh, P. R. Kumar, and S. N. Mohanty, "A manifold-level hybrid deep learning approach for sentiment classification using an autoregressive model," *Applied Sciences*, vol. 13, no. 5, p. 3091, 2023.
8. L. Wu and J. Du, "Designing novel proxy-based access control scheme for implantable medical devices," *Computer Standards & Interfaces*, vol. 87, p. 103754, 2024, doi: 10.1016/j.csi.2023.103754.
9. M. Mahdavi, M. H. Tadayon, M. S. Haghighi, and Z. Ahmadian, "IoT-friendly, pre-computed and outsourced attribute based encryption," *Future Generation Computer Systems*, vol. 150, pp. 115–126, 2024, doi: 10.1016/j.future.2023.08.015.

10. N. Arivazhagan, K. Somasundaram, G. B. Mohammad, P. R. Kumar, et al., "Cloud-Internet of Health Things (IOHT) task scheduling using hybrid moth flame optimization with deep neural network algorithm for e-healthcare systems," *Scientific Programming*, vol. 2022, pp. 1–12, 2022.
11. F. Meng, "Online/offline attribute-based searchable encryption revised: Flexibility, security and efficiency," *Journal of Systems Architecture*, vol. 146, p. 103047, 2024, doi: 10.1016/j.sysarc.2023.103047.
12. P. R. Kumar, G. B. Mohammad, and P. Dileep, "Real-time heart rate monitoring system using least square method," *Annals of the Romanian Society for Cell Biology*, vol. 25, no. 6, pp. 16302–16308, 2021.
13. P. R. Kumar, and T. Ananthan, "Machine vision using LabVIEW for label inspection," *Journal of Innovation in Computer Science and Engineering (JICSE)*, vol. 9, no. 1, pp. 58–62, 2019.
14. J. Lee, J. Oh, D. Kwon, M. Kim, K. Kim, and Y. Park, "Blockchain-enabled key aggregate searchable encryption scheme for personal health record sharing with multi-delegation," *IEEE Internet of Things Journal*, pp. 1–1, 2024, doi: 10.1109/jiot.2024.3357802.
15. P. R. Kumar, "Wireless mobile charger using inductive coupling," *Journal of Emerging Technologies and Innovative Research (JETIR)*, vol. 5, no. 10, pp. 40–44, 2018.
16. S. Ilakiya, "Literature survey on Attribute Based Encryption (ABE)," *International Journal of Emerging Trends in Science and Technology*, 2016, https://journals.indexcopernicus.com/api/file/viewByFileId/182258.
17. M.-S. Lacharité, "Security of BLS and BGLS signatures in a multi-user setting," *Cryptography and Communications*, vol. 10, no. 1, pp. 41–58, 2017, doi: 10.1007/s12095-017-0253-6.
18. J. Mary and S. Sunitha, "ABE based online personal health record system using cloud computing," *International Journal of Innovative Research in Computer and Communication Engineering*, vol. 3, no. 7, pp. 6818–6823, 2015, doi: 10.15680/ijircce.2015.0307035.
19. Y. Zhao, Y. Fan, and X. Bian, "OO-MA-KP-ABE-CRF: Online/offline multi-authority key-policy attribute-based encryption with cryptographic reverse firewall for physical ability data," *Mathematics*, vol. 11, no. 15, p. 3333, 2023, doi: 10.3390/math11153333.
20. B. Shilpa, P. R. Kumar, and R. K. Jha, "Spreading factor optimization for interference mitigation in dense indoor LoRa networks," *IEEE IAS Global Conference on Emerging Technologies (GlobConET)*, London, UK pp. 1–5, 2023.
21. J. Jusak, S. S. Mahmoud, R. Laurens, M. Alsulami, and Q. Fang, "A new approach for secure cloud-based electronic health record and its experimental testbed," *IEEE Access*, vol. 10, pp. 1082–1095, 2022, doi: 10.1109/access.2021.3138135.
22. S. Senthilkumar, K. Brindha, N. Kryvinska, S. Bhattacharya, and G. R. Bojja, "SCB-HC-ECC-based privacy safeguard protocol for secure cloud storage of smart card–based health care system," *Frontiers in Public Health*, vol. 9, 2021, doi: 10.3389/fpubh.2021.688399.
23. P. R. Kumar, "Position control of a stepper motor using LabVIEW," *3rd International Conference on Recent Trends in Electronics, Information Communication Technology (RTEICT)*, Bangalore, India, pp. 1551–1554, May 2018.

24. P. R. Kumar and B. Shilpa, "An IoT-based smart healthcare system with edge intelligence computing," In S. Satpathy, S. N. Mohanty, and S. Potluri (eds.), *Reconnoitering the Landscape of Edge Intelligence in Healthcare.* CRC Press, Boca Raton, FL, pp. 31–46, 2024.
25. R. Ganiga, R. M. Pai, M. M. Manohara Pai, and R. K. Sinha, "Security framework for cloud based electronic health record (EHR) system," *International Journal of Electrical and Computer Engineering (IJECE)*, vol. 10, no. 1, pp. 455–466, 2020, doi: 10.11591/ijece.v10i1.pp455-466.

Chapter 6

Machine learning unleashed

A paradigm shift in blockchain intelligence

Marepalli Radha, MD Asma, Ashlesha Kolarkar, and Kuncham Sreenivasa Rao

6.1 INTRODUCTION

Innovative applications, including connected health care, smart cities, and connected industries, have been developed as a result of the rapid development of emerging technologies, smartphones, sensors, and 5G communication. This has contributed to the production of enormous quantities of data. There are legitimate security issues regarding the estimated 50.1% increase in the number of Internet of Things (IoT) devices linked to the Internet by 2020, according to a study conducted by the National Cable & Telecommunications Association (NCTA) [1]. Cyberattacks and data breaches, particularly targeting IoT devices, have surged, with McAfee reporting a barrage of incidents across various industries since January 2018. The vulnerability of IoT devices is exacerbated by their increasing interconnectivity, as highlighted by VDC Research Group Inc., which identifies security requirements as a significant obstacle in developing connected devices [2]. Kaspersky Lab's data reveals a substantial rise in malware samples for IoT devices from 2016 to 2018, underscoring the substantial vulnerabilities in these devices. Monitoring network-based risks presents substantial issues for a variety of industries, including the government, energy, healthcare, banking, and research centres, among others. The enormous amount, speed, variety, and authenticity of data makes current methods and technologies insufficient to identify novel cyberattacks on IoT devices [3]. If you are dealing with massive amounts of data, weekly or monthly security analytics reports won't cut it. The research acknowledges the shortcomings and suggests combining deep learning with big data to strengthen the security of IoT devices. Deep learning is well-suited for networks with limited resources because of its compression characteristics, ability to train without supervision, and lack of human intervention in feature building. This study intends to thoroughly investigate the practicability of integrating these technologies, despite the fact that deep learning, big data, and IoT security have all received independent attention in the current literature. By evaluating data flow to discover intrusions and attack

 DOI: 10.1201/9781032663005-6

patterns, the deep extreme learning machine (DELM) tackles the standard signature-based technique in Intrusion Detection System (IDS) [4].

The integration of smart blockchain-based applications and robust algorithms for data processing is essential to manage the substantial data generated by IoT devices. Machine learning, as part of artificial intelligence (AI) frameworks, involves training machines to make predictions and decisions based on statistical analysis [5]. This research advocates for the use of DELM to enhance the security of IoT-enabled smart homes, providing a thorough review of cutting-edge technologies and proposing an architecture for implementation in blockchain-based smart homes. The DELM framework, in conjunction with blockchain, offers a unique solution for various applications, including fraud detection and theft prediction, by focusing on specific chain segments and eliminating data-related issues.

The organization of this article unfolds as follows: The subsequent section provides a concise summary of relevant survey articles. We then delve into the foundational framework of blockchain, elucidate the application of the DELM approach within the context of blockchain-based smart homes, and delineate the framework for smart home applications. Subsequently, we examine the simulation and testing procedures employed for the DELM approach. The final section of the article delves into the research conclusions.

6.2 LITERATURE SURVEY

This segment offers an in-depth depiction of deep learning, big data technologies, and IoT security. Moreover, it explores the interconnections between these three realms, aiming to furnish essential insights and a mapping of the relationships among these cutting-edge subjects.

6.2.1 Deep learning

Deep Learning, a subset of machine learning, encompasses three distinct learning techniques: supervised, semi-supervised, and unsupervised learning. It is characterized by the presence of multiple layers in artificial neural networks (ANNs), each comprising neurons with activation functions capable of generating non-linear outputs. This approach draws inspiration from the structure of neurons in the human brain [6]. Over recent years, deep learning has garnered significant attention from researchers and organizations, surpassing the interest in traditional machine learning methods.

Deep learning was compared to four other machine learning methods, namely Support Vector Machine (SVM), Decision Trees, K-means, and Logistic Regression, by the authors of Ref. [7]. They used Google Trends to carry out their investigation. The results point to deep learning's rising profile. Natural language processing, search engines, information retrieval, picture identification, and image retrieval are just a few of the AI disciplines

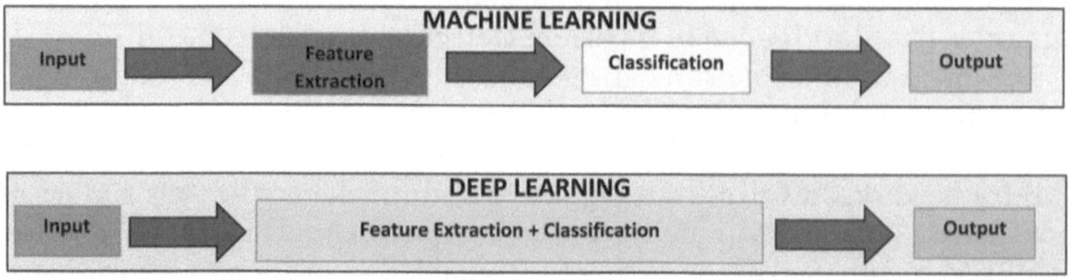

Figure 6.1 Machine learning vs. deep learning.

that have made use of this technology. There are four steps to building a model in both deep learning and machine learning. Figure 6.1 shows how deep learning differs from machine learning. In this section, we will go over the fundamentals of deep learning before diving into its common techniques and traits.

6.2.2 Big data technologies

Big data describes the information that requires new ways of processing to gain insights and make decisions due to its large volume, high velocity, and diversity [8]. Figure 6.3 depicts the six defining features, or "the 6Vs," that are usually associated with big data. The first three V's—volume, velocity, and variety—are requirements for data to be considered big data, but. When it comes to processing large amounts of data efficiently, the tools and technologies used are known as big data technology. Apache Hadoop [9], Apache Spark [10], Apache Storm [11], Apache Flink [12], Apache Cassandra [13], and Apache HBase [14] are all examples of such technologies. Earlier, we listed some of the most popular big data technologies and described the characteristics of big data, which include the 6V's.

With the help of the IoT, smart home sensors and other gadgets may talk to one other and share data across different systems. Smart cities, homes, offices, retail outlets, agriculture, water management, transportation, healthcare, and energy are just a few examples of the intelligent systems that have recently benefited from the extensive use of the IoT [15]. Data capture in the IoT world is facilitated by mobile devices, transportation facilities, public areas, and domestic appliances, all of which are heavily reliant on the IoT. In addition, the IoT network allows for the remote control of devices across different applications, which in turn allows them to communicate with one other and with central controlling devices. The IoT enables the gathering of geographical, astronomical, environmental, and logistical data, among others, across a wide range of disciplines [16]. Protecting the entire IoT deployment architecture from possible assaults is what IoT security is all about [17]. There are a number of aspects to think about while developing security solutions for the IoT. There are several factors to think about while developing

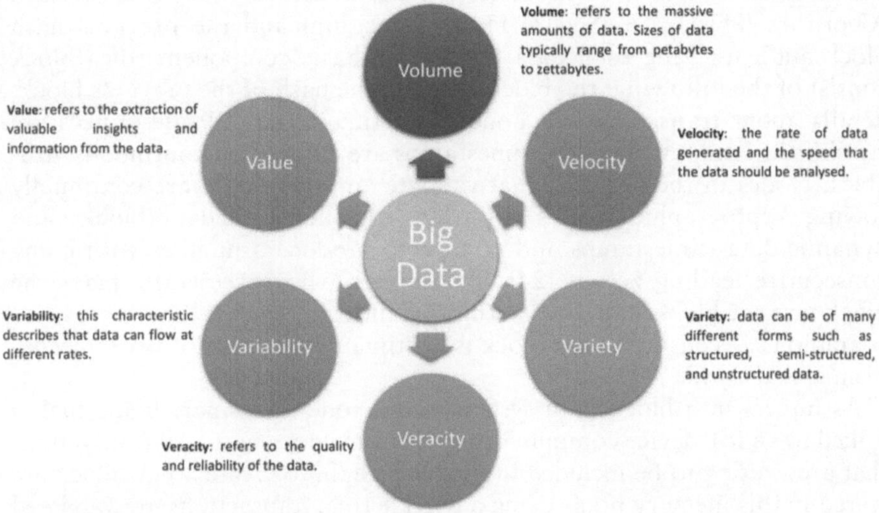

Figure 6.2 Six Vs of big data.

strong IoT security solutions. We can detect a spectrum of security breaches linked with these criteria by leveraging the capabilities of deep learning and big data technology. Figure 6.2 represents 6Vs of the big data.

Confidentiality plays a crucial role in ensuring secure information transmission in all communications. When information is transmitted without proper authentication or encryption, it exposes the possibility of privacy violations by adversaries [18]. In the context of big data technologies, secure data transmission is typically achieved through encryption methodologies, thereby preventing unauthorized access and compromise of data by adversaries [19]. Integrity is vital for preserving the trustworthiness of an IoT system, as adversaries may attempt to compromise it. The data received has not been altered during transmission if integrity is maintained [20]. It is worth mentioning that Apache Spark, a big data platform, allows users to do integrity checks on the IoT system by supporting data quality checks within the Spark DataFrame [21]. When discussing IoT systems, "availability" means keeping the system accessible to authorized users while blocking access to unauthorized ones [22]. This objective is in line with big data technologies, which prioritize their accessibility to users and their capacity to run on numerous nodes, guaranteeing that applications will be highly available [23].

6.3 METHODOLOGY

Simple blockchain cryptocurrencies like bitcoin were introduced by Satoshi Nakamoto in 2008 as part of a peer-to-peer payment network that aimed to eliminate middlemen and solve the problem of double spending.

The system functions as a clustered data structure, with Secure Hash Algorithm (SHA)-256 (Secure Hash Algorithm) and the previous hash block authenticating each data block. The basic components of a block consist of the following: the block number, the hash of the previous block, details about transactions, a nonce, and timestamps. While nonces are considered random variables, timestamps are considered continuous variables. Nodes in the network that validate and mine data are continually solving cryptographic puzzles by hashing together static data (blocks) and dynamic data (timestamps and nonces) to produce a number with many consecutive leading zeroes [24]. The miners who successfully place the block into the blockchain by determining the right hash value are crowned champions. To ensure that a block is legitimate, the proof of work mechanism is employed.

As miners in a blockchain system, every node in a smart home that is linked to an IoT device communicates with a memory pool. All transactions that are waiting to be included in the blockchain to create a new block are stored in this memory pool. Using a Merkle tree, transactions are validated and summarized. Then, miners all throughout the smart home system will be able to see the valid transactions that have been added to the block. To create a Hash of Block, miners tweak the nonce and timestamp. After that, the programme tries to check if the generated hash matches the target [25]. The hash is attached to the chain as soon as a miner produces a valid block. This procedure repeats until the hash value is greater than or equal to the specified threshold. The proof of work is added to the chain and validated for efficiency if the hash value is smaller than the target value. This message is sent out to all nodes in the network to let them know that the memo pool transactions are complete. Because of its adaptability and interoperability with smart home IoT applications, blockchain technology is having an ever-increasing impact on the smart home communication environment [26]. The four-layer blockchain-based smart home network shown in Figure 6.1 consists of an IoT data source layer, a blockchain network layer with DELM capabilities, a layer for intelligent home devices, and a client node.

6.3.1 Integration of deep extreme learning machine in blockchain-based smart home

In 2008, blockchain was introduced by M. C. Thrun et al. [27], providing a peer-to-peer payments network that addresses third-party removal and double-spending issues through a simple blockchain cryptocurrency like bitcoins. This system operates as a clustered data structure, with each data block authenticated by SHA-256 along with the previous hash block. A basic block structure has a block number, a hash of the previous block, transaction information, a nonce, and timestamps. The goal of validators and miners in solving cryptographic puzzles, which involve hashing both static and dynamic data, is to find a number with consecutive leading zeroes. Winners are the miners who correctly guess the hash value and are given the green light to insert the block, which is then validated using the proof of work mechanism. Every IoT device in a smart home communicates with a shared memory pool in the same way as blockchain miners do their work. All of the transactions that will be part of the next block on the blockchain are stored in this memory pool. Using a Merkle tree, transactions are validated and summarized [28]. Then, miners all throughout the smart home system will be able to see the valid transactions that have been added to the block. To create a Hash of Block, miners tweak the nonce and timestamp. After that, the programme tries to check if the generated hash matches the target. The hash is attached to the chain as soon as a miner produces a valid block. This procedure repeats until the hash value is greater than or equal to the specified threshold. The proof of work is added to the chain and validated for efficiency if the hash value is smaller than the target value. This message is sent out to all nodes in the network to let them know that the memo pool transactions are complete.

Because of its adaptability and interoperability with smart home IoT applications, blockchain technology is having an ever-increasing impact on the smart home communication environment. The four-layer blockchain-based smart home network shown in Figure 6.1 consists of an IoT data source layer, a blockchain network layer with DELM capabilities, a layer for intelligent home devices, and a client node. Critical data for assessing smart homes, surroundings, and users is collected by the IoT information layer from various devices. Sensors, multimedia, and medical devices are the three main types of these gadgets. Thermostats and other sensors in the IoT sensor network determine and control environmental factors, while closed-circuit television and wearables are part of the same network. The first layer of the stack consists of central databases or repositories that gather information from these nodes; one example is a blockchain. Blockchain-based apps can be made smarter by applying DELM computing technology. Distributed ledger security and information sharing route efficiency can both be improved using DELM. Additionally, it opens the door to possibilities for enhancing frameworks through the use of blockchain technology's centralized design.

By running the datasets utilized by DELM models across a blockchain network, the suggested DELM framework is able to lessen the impact of mistakes such as duplicate data, missing values, flaws, and noise. Offering distinctive frameworks for a range of applications, including fraud detection, the DELM framework can zero down on particular chain pieces instead of the full dataset [29]. Smart contracts, the DELM layer, and the blockchain information architecture are the three main components of the model that is built on top of the IoT edge architecture. Included in the smart home framework are technologies like digital markets, access control, healthcare-home integration, intelligent community services, and automated infrastructure payments. At the very top of the hierarchy is the access layer, which opens the door for third parties like microgrids, retailers, and utility companies to use smart home devices built on the blockchain.

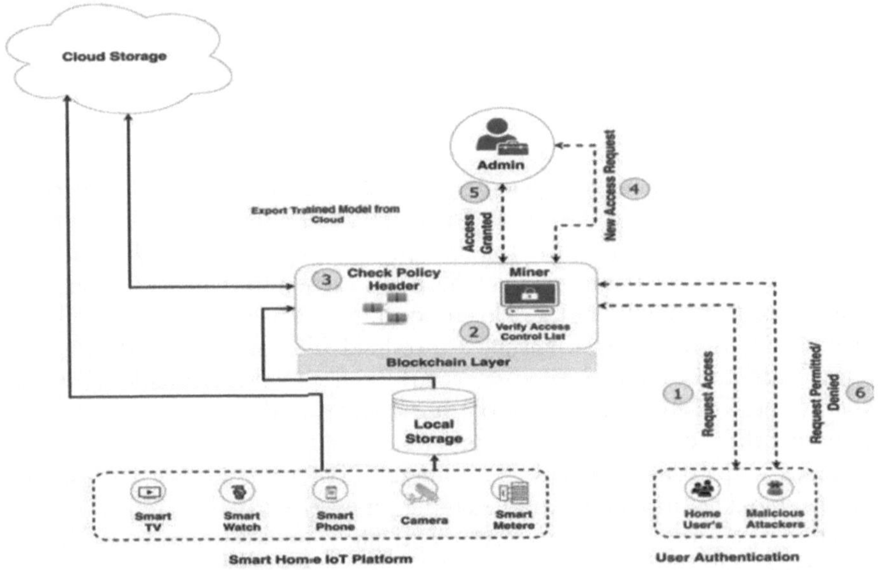

Figure 6.3 Blockchain-based smart home management system empowered with deep extreme learning machine.

The backbone of improved smart home environments are smart home gadgets including cameras, CCTV, smart TVs, fitness trackers, cellphones, and actuators. Features such as remote control, alarm generating, and safety surveillance are provided by these devices. In order to keep smart home operations running smoothly and identify any suspicious activity, a personalized access control system is essential. A collection of immutable distributed information network access records for every user contains access permission specifications for the IoT system [30].

Using a home user (Admin) with individualized access to the smart house and its applications as an example, we can see how blockchain guarantees secure access. Secure access is made possible by blockchain, as seen in Figure 6.3. After determining the appropriate level of access, users must assign it to the home service machine. Homeowners (as Admins) have full power, whereas minors, guests, and strangers have intermediate or lower-level permissions. Whenever a user requests legal authentication, the home server checks the access control directory and then notifies the blockchain layer. The authorization list for various users and gadgets is stored in a blockchain policy header. After an administrator approves or rejects an access request, blockchain miners incorporate the policy details into the header and take action, protecting the network from harmful attacks.

6.3.2 Deep extreme learning machine

The DELM is useful in many areas, such as predicting health problems, figuring out how much energy a building will use, transportation, traffic control, and more [1]. As described by Ref. [13], the DELM stands out because to its fast-learning capability and efficacy in procedural convolution rates, in contrast to existing ANN algorithms that frequently necessitate several changes and lengthy learning cycles. Because of its fast learning and procedural convolution rate, the DELM is relevant and useful in many domains for regression and classification purposes.

As a feedforward neural network, the XLR usually only permits bidirectional data flow over its many layers. But when it comes time to learn, the suggested system uses a backpropagation method. By using this method, data may be fed back into the network, letting the neural network fine-tune its weights for optimal performance with little room for error. During validation, when the trained model is used to make predictions based on real data, the weights stay the same. An input layer, two or more hidden layers, and an output layer make up the suggested DELM method. A DELM has numerous hidden layers with a fixed number of neurons, as opposed to the single hidden layer with multiple neurons that is typical in extreme learning machines. As a result of this setup, the network is more efficient. As seen in Table 6.1, we outperformed previous machine learning methods by adding more hidden layers while keeping the number of neurons constant.

To minimize the error rate and change the network weights, DELM combines the backpropagation and feedforward techniques. When contrasted with other machine learning techniques, the DELM framework proves to be more accurate. In order to optimize smart home security, the assessment layer observes many statistical factors, as shown in Figure 6.4. Precision, Sensitivity, Specificity, Positive Prediction Value, and False Positive Rate are some of the metrics that fall under this category. Configuring weights, feeding them forwards, propagating backward errors, and updating

Table 6.1 Comparison of the proposed DELM method with other machine learning algorithms with different datasets

Method	*Accuracy of Network Security Laboratory - Knowledge Discovery and Data Mining (NSL-KDD) dataset [14] (%)*	*Accuracy of Knowledge Discovery and Data Mining Cup (KDD-CUP-99) dataset [15] (%)*
Artificial Neural Network (ANN)	82.3	91.49
Support Vector Machine (SVM)	68.41	88.44
Decision Tree (DT)	82.7	92.24
Proposed DELM	94.87	95.7

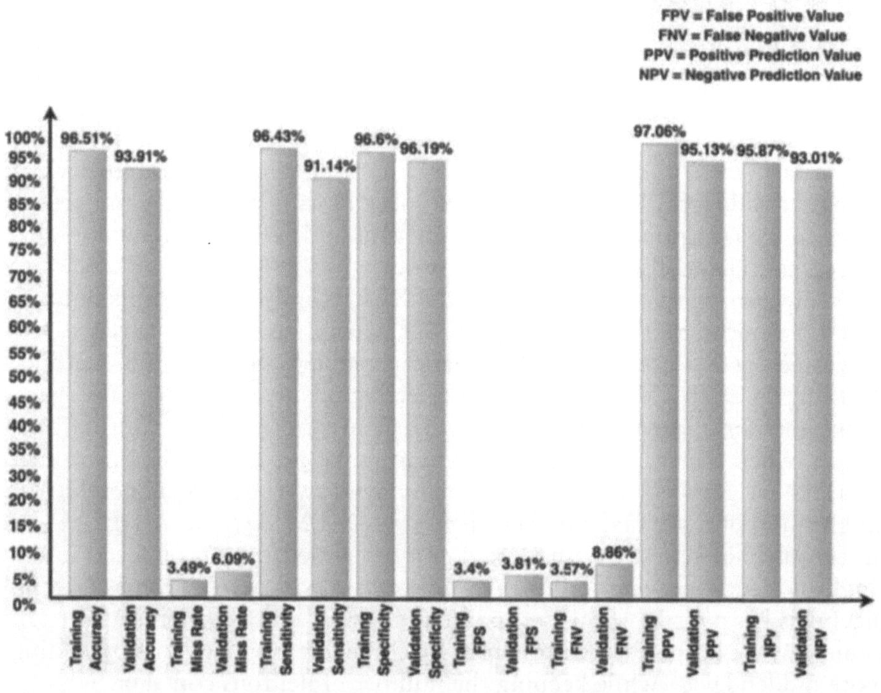

Figure 6.4 Performance evaluation of blockchain-based smart home empowered with deep extreme learning machine system model during the prediction of malicious activity or attacks using different statistical measure.

distinguishability are all parts of the backpropagation process. The design of the DELM hidden layer and the sigmoid input function are influenced by the fact that each hidden layer neuron is subjected to a sigmoid activation function. If the squared difference between the output and the input is less than two, then the hidden layer is doing well. To reduce frequent mistakes in the network, weight modifications are crucial.

6.4 RESULTS AND DISCUSSION

In this study, the proposed framework implemented the DELM using input data from Ref. [14]. The dataset was randomly partitioned into 85% for training (125,973 samples), and the remaining 15% was allocated for validation (22,543 samples). Prior preprocessing of the data was conducted to eliminate irregularities and minimize the potential impact of information errors. DELM was employed to detect malicious activities or intrusions across various hidden layers, hidden connections, and activation functions. The evaluation involved assessing different numbers of neurons in hidden layers and various types of activation functions. The efficiency of the DELM system was properly evaluated in this analysis. To compare the output with other algorithms, various statistical measures were employed. Table 6.2 presents the intrusion detection model predictions during the training phase for the proposed blockchain-based smart home system empowered by the DELM. The training dataset comprised a total of 125,973 samples, with 67,343 samples representing normal instances and 58,630 samples representing attacks.

During the analysis, it was found that 66,423 samples from the normal class (indicating instances with no detected attacks) were accurately predicted, while 1,867 records were erroneously predicted as an attack, despite no actual attack occurring. Similarly, in the case of detected attacks, out of a total of 58,630 samples, 57,620 were correctly identified as attacks, while 2,320 samples were inaccurately predicted as normal instances, even though an attack was present. Table 6.3 illustrates the intrusion detection model predictions for the proposed blockchain-based smart home system empowered by the DELM during the validation phase. The validation dataset consisted of 22,543 samples, with 9,610 representing normal instances and 12,733 representing attacks.

Furthermore, additional statistical measures have been incorporated to predict values, including false positives, false negatives, likelihood ratios (negative and positive), as well as positive and negative prediction values.

Table 6.2 Predicting harmful actions or assaults and training a model for a blockchain-enabled smart home using deep extreme learning

Proposed DELM-based system model (85% of sample data in training)			
Total number of samples (N= 125,973)		*Output results (O_0, O_1)*	
Input	*Expected output (T_0, T_1)*	*O_0 (normal)*	*O_1 (attack)*
	T_0=66,423 normal	65,256	1,867
	T_1=57,620 attack	2,320	56,120

Table 6.3 Predicting harmful actions or assaults using a blockchain-based smart home equipped with a deep extreme learning machine system model

Proposed DELM-based system model (15% of sample data in validation)			
Total number of samples (N=22,443)		*Output results (O_0, O_1)*	
Input	*Expected output (T_0, T_1)*	*O_0 (normal)*	*O_1 (attack)*
	T_0=9,610 normal	9,337	463
	T_1=12,733 attack	878	11,835

The outcomes of these measures are presented in Figure 6.4. The analysis reveals that 9,237 samples from the normal class were accurately predicted, while 473 records were erroneously predicted as an attack despite the absence of any actual attack. Similarly, in the case of detected attacks, out of a total of 12,733 samples, 11,935 were correctly identified as attacks, while 898 samples were inaccurately predicted as normal instances when an attack was present. Figure 6.3 illustrates the performance of the proposed blockchain-based smart home system empowered by the DELM in terms of various statistical measures during both the training and validation phases. The results clearly indicate that during training, the proposed system achieves an accuracy of 95.7% and a miss rate of 3.49%.

In the validation phase, the proposed system achieves an accuracy of 94.87% and a miss rate of 6.09%. The system model performance is also depicted in terms of sensitivity and specificity during both the training and testing phases. The results indicate that during training, the proposed system achieves 96.43% sensitivity and 96.6% specificity, while during validation, it achieves 91.14% sensitivity and 96.19% specificity.

6.5 CONCLUSION

In particular, evaluation and prediction present formidable obstacles to the detection of intrusions in smart homes. New developments in blockchain technology and AI hold great potential for tackling these issues. Yet, it is difficult to execute such solutions efficiently due to the power and processing limits of devices in the majority of smart home configurations. This research sought to address this knowledge vacuum by presenting a minimally invasive approach to intrusion prediction and detection that makes use of a blockchain-based architecture and DELM. The proposed solution was evaluated using several statistical approaches; the findings show that the DELM method is more reliable than the others. With an impressive accuracy rate of 94.81%, the suggested DELM method produced outstanding results. Additional datasets and various architectures are being considered as part of the ongoing investigation into these promising results.

REFERENCES

1. M. Wazid, P. Bagga, A. K. Das, S. Shetty, J. J. P. C. Rodrigues, and Y. H. Park, "AKM-IoV: Authenticated key management protocol in fog computing-based internet of vehicles deployment," *IEEE Internet of Things Journal*, vol. 6, no. 5, pp. 8804–8817, 2019.
2. B. Shilpa, P. R. Kumar, and R. K. Jha, "LoRa DL: A deep learning model for enhancing the data transmission over LoRa using autoencoder," *The Journal of Supercomputing*, vol. 79, pp. 17079–17097, 2023.

3. L. Nie, Z. Ning, X. Wang, X. Hu, J. Cheng, and Y. Li, "Datadriven intrusion detection for intelligent internet of vehicles: A deep convolutional neural network-based method," *IEEE Transactions on Network Science and Engineering*, vol. 7, no. 4, pp. 2219–2230, 2020.
4. Q. Ali, N. Ahmad, A. Malik, G. Ali, and W. Rehman, "Issues, challenges, and research opportunities in intelligent transport system for security and privacy," *Applied Sciences*, vol. 8, no. 10, p. 1964, 2018.
5. R. Ranjan, D. Pandey, A. K. Rai, D. Gupta, P. Singh, P. R. Kumar, and S. N. Mohanty, "A manifold-level hybrid deep learning approach for sentiment classification using an autoregressive model," *Applied Sciences*, vol. 13, no. 5, p. 3091, 2023.
6. S. Garg, A. Singh, G. S. Aujla, S. Kaur, S. Batra, and N. Kumar, "A probabilistic data structures-based anomaly detection scheme for software-defined internet of vehicles," *IEEE Transactions on Intelligent Transportation Systems*, vol. 22, no. 6, pp. 3557–3566, 2020.
7. N. Arivazhagan, K. Somasundaram, G. B. Mohammad, P. R. Kumar, et al., "Cloud-Internet of Health Things (IOHT) task scheduling using hybrid moth flame optimization with deep neural network algorithm for e-healthcare systems," *Scientific Programming*, vol. 2022, pp. 1–12, 2022.
8. M. Keshk, N. Moustafa, E. Sitnikova, and B. Turnbull, "Privacy preserving big data analytics for cyber-physical systems," *Wireless Network*, 28, pp. 1–9, 2018.
9. P. Kumar, et al., "PPSF: A privacy-preserving and secure framework using blockchain-based machine-learning for IoT-driven smart cities," *IEEE Transactions on Network Science and Engineering,* 2021, doi: 10.1109/TNSE.2021.3089435.
10. G. B. Mohammad, Selvarajan Shitharth, and P. R. Kumar, "Integrated machine learning model for an URL phishing detection," *International Journal of Grid and Distributed Computing*, vol. 14, no. 1, pp. 513–529, 2021.
11. H. Gao, W. Huang, and X. Yang, "Applying probabilistic model checking to path planning in an intelligent transportation system using mobility trajectories and their statistical data," *Intelligent Automation & Soft Computing*, vol. 25, no. 3, pp. 547–559, 2019.
12. M. B. Mollah et al., "Blockchain for the internet of vehicles towards intelligent transportation systems: A survey," *IEEE Internet of Things Journal*, vol. 8, no. 6, pp. 4157–4185, 2021.
13. P. R. Kumar and T. Ananthan, "Machine vision using LabVIEW for label inspection," *Journal of Innovation in Computer Science and Engineering (JICSE)*, vol. 9, no. 1, pp. 58–62, 2019.
14. R. Kumar and R. Tripathi, "DBTP2SF: A deep blockchain-based trustworthy privacy-preserving secured framework in industrial Internet of Things systems," *Transactions on Emerging Telecommunications*, vol. 32, no. 4, p. e4222, 2021.
15. S. Singh, P. K. Sharma, B. Yoon, M. Shojafar, G. H. Cho, and I.-H. Ra, "Convergence of blockchain and artificial intelligence in IoT network for the sustainable smart city," *Sustainable Cities and Society*, vol. 63, p. 102364, 2020.
16. P. R. Kumar, "Wireless mobile charger using inductive coupling," *Journal of Emerging Technologies and Innovative Research (JETIR)*, vol. 5, no. 10, pp. 40–44, 2018.

17. T. Rausch and S. Dustdar, "Edge intelligence: The convergence of humans, things, and AI," In *2019 IEEE International Conference on Cloud Engineering (IC2E)*. IEEE, Prague, Czech Republic pp. 86–96, 2019.
18. P. Kairouz, H. B. McMahan, B. Avent, A. Bellet, M. Bennis, A. N. Bhagoji, K. Bonawitz, Z. Charles, G. Cormode, R. Cummings, et al., "Advances and open problems in federated learning," arXiv preprint arXiv:1912.04977, 2019.
19. B. Shilpa, P. R. Kumar, and R. K. Jha, "Spreading factor optimization for interference mitigation in dense indoor LoRa networks," *IEEE IAS Global Conference on Emerging Technologies (GlobConET)*, London, UK pp. 1–5, 2023.
20. Q. Yang, Y. Liu, T. Chen, and Y. Tong, "Federated machine learning: Concept and applications," *ACM Transactions on Intelligent Systems and Technology (TIST)*, vol. 10, no. 2, pp. 1–19, 2019.
21. S. Georgi and R. Jung, "Collective intelligence model: How to de- scribe collective intelligence," In J. Altmann, U. Baumöl, and B. Krämer (eds.), *Advances in Collective Intelligence 2011*. Springer, New York, pp. 53–64, 2012.
22. J. H. Watkins, "Prediction markets as an aggregation mechanism for collective intelligence," *Proceedings of the Human Complex System Conference*, Lake Arrowhead, CA, 2007.
23. R. M. Livingstone, "Models for understanding collective intelligence on Wikipedia," *Social Science Computer Review*, vol. 34, no. 4, pp. 497–508, 2016.
24. K. Zettsu and Y. Kiyoki, "Towards knowledge management based on harnessing collective intelligence on the web," In *International Conference on Knowledge Engineering and Knowledge Management*. Springer, Nancy, France pp. 350–357, 2006.
25. P. R. Kumar, "Position control of a stepper motor using LabVIEW," *3rd International Conference on Recent Trends in Electronics, Information Communication Technology (RTEICT)*, Bangalore, India pp. 1551–1554, May 2018.
26. S. Zhang, Y. Chen, W. Zhang, and R. Feng, "A novel ensemble deep learning model with dynamic error correction and multi-objective ensemble pruning for time series forecasting," *Information Sciences*, vol. 544, pp. 427–445, 2021.
27. M. C. Thrun and A. Ultsch, "Swarm intelligence for self-organized clustering," *Artificial Intelligence*, vol. 290, pp. 103237, 2021.
28. M. Rafiee and A. M. Bayen, "Optimal network topology design in multiagent systems for efficient average consensus," In *49th IEEE Conference on Decision and Control (CDC). IEEE,* Atlanta, GA, USA pp. 3877–3883, 2010.
29. P. R. Kumar and B. Shilpa, "An IoT-based smart healthcare system with edge intelligence computing," In S. Satpathy, S. N. Mohanty, and S. Potluri (eds.), *Reconnoitering the Landscape of Edge Intelligence in Healthcare*. CRC Press, Boca Raton, FL, pp. 31–46, 2024.
30. V. D. Nguyen and N. T. Nguyen, "Intelligent collectives: Theory, applications, and research challenges," *Cybernetics and Systems*, vol. 49, no. 5–6, pp. 261–279, 2018.

Chapter 7

Securing HPC data clusters with in-memory blockchain

A provenance enhancement approach

Maradana Durga Venkata Prasad and Srikanth T

7.1 INTRODUCTION

Blockchain represents a distributed storage paradigm that incorporates various technologies, including encryption algorithms, peer-to-peer networks, and consensus algorithms [1]. Recognized as a revolutionary innovation, its decentralization, tamper-proof information, and traceability features have made it a disruptive force in recent years. Despite its transformative potential, large-scale commercial blockchain applications demand a high-performance and scalable storage architecture, surpassing the capabilities of traditional public blockchain architectures [2]. This has led to the emergence of consortium blockchains, specifically designed to handle extensive transactions in business scenarios, offering superior storage performance compared to public counterparts [3].

However, existing consortium blockchain platforms face storage bottlenecks, primarily rooted in their reliance on key-value databases like LevelDB. As data volumes increase, this results in significant reading/writing amplification and constant compaction, diminishing storage efficiency and causing performance bottlenecks. Furthermore, key-value databases prove unsuitable for handling substantial single pieces of data, leading to exponential increases in storage delays as transaction data size grows. Efforts to enhance consortium blockchain storage performance have mainly focused on reducing data storage volume rather than restructuring the underlying architecture [4]. Approaches such as compressing node data and collaborative data storage aim to minimize data in blockchain nodes. However, excessive data reduction risks compromising the integrity of blockchain records. Some researchers propose distributed extensions, particularly employing distributed file systems, to address massive data storage challenges. Yet, these architectures prioritize storage extension over performance, failing to meet the efficiency needs of enterprise-level business scenarios requiring rapid and reliable reading/writing capabilities (Figure 7.1).

Provenance systems traditionally fall into two categories: centralized and distributed. SPADE, a prominent centralized system, manages provenance

DOI: 10.1201/9781032663005-7

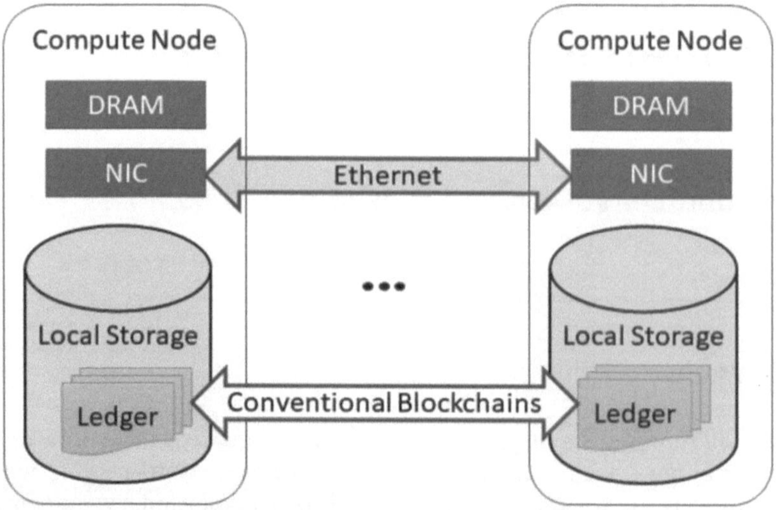

Figure 7.1 Conventional blockchain architecture on shared-nothing platforms.

collected from various sources using a centralized relational database management system (RDBMS). Domain-specific variations following similar centralized design principles are prevalent in fields like biomedical engineering and computational chemistry [5]. However, the exponential growth of data has led to criticism of centralized systems. They become performance bottlenecks and single points of failure. This has prompted the development of distributed approaches for scalable provenance. Distributed provenance systems, often built on distributed file systems rather than centralized databases, address the performance limitations of centralized counterparts. They demonstrate significantly higher performance, but this shift introduces a new concern: who should audit the provenance in distributed systems? This question raises the issue of building the provenance of provenance, creating an endless recursion. While this concern was less critical in centralized approaches due to the application of robust reliability mechanisms to a centralized node, it poses a significant challenge in large-scale distributed systems [6]. In such a setup, if any single node is compromised, the entire provenance becomes invalid. To tackle this challenge, recent advancements in distributed provenance systems draw inspiration from blockchain technology (Figure 7.2).

In order to keep track of where data comes from in high-performance computing (HPC) systems, this study presents a new blockchain architecture. The architecture has been meticulously crafted to conform to the specific requirements of the HPC environment. Importantly, compute nodes can choose to keep the blockchain running even when they do not have access to local discs. Reduced persistent data size and I/O overhead are the

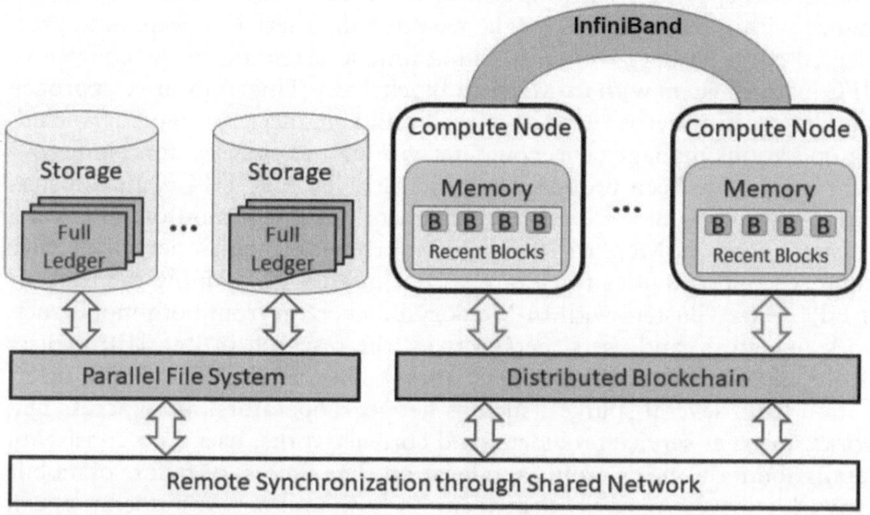

Figure 7.2 Proposed high-performance computing blockchain architecture with shared storage.

outcomes of this novel technique. To verify the origin of apps' data, the article suggests a new consensus mechanism called Proof-of-Reproducibility (PoR) in addition to the tailored architecture [7]. The underlying idea behind Proof-of-Relay (PoR) is that consensus may be reached by utilizing both Proof-of-Work (PoW) and Proof-of-Stake (PoS), which are distributed ledger technologies (DLT). This novel consensus protocol's soundness is formally proven. A system prototype is developed using around 1,800 lines of Java code in order to verify the suggested ideas [8]. We demonstrate the practicality of the proposed blockchain architecture and consensus protocol in the context of HPC systems by conducting experimental verification of the system's effectiveness.

In the subsequent sections of this work, we will examine relevant literature in Section 7.2. Section 7.3 presents a blockchain architecture specifically designed for HPC. The implementation details of a prototype for the proposed blockchain system are elaborated upon in Section 7.4. The experimental findings are outlined within the same section. The chapter is concluded in Section 7.5.

7.2 RELATED WORK

This section delves into the scientific research associated with our exploration of enhancing the security of HPC Data Clusters through In-Memory Blockchain. As outlined in Ref. [9], a methodology is presented for querying provenance, specifically distinguishing between "where" and "why" provenance within databases. Over the past decade, there has been an unprecedented surge in data traffic, bringing unique attention to the concept of HPC Data Clusters with In-Memory Blockchain. This innovative approach is under scrutiny across various scientific and engineering domains, including operations management, computer vision, and smart cities. Numerous techniques have been proposed to efficiently leverage HPC Data Clusters with In-Memory Blockchain. For instance, the integration of HPC Data Clusters with In-Memory Blockchain provides mobile networks with enhanced opportunities to elevate service quality. Examining the features of HPC Data Clusters with In-Memory Blockchain from both mobile network operators' and users' perspectives, the research in Ref. [10] focuses on the incorporation of mobile networks. This requires combining information from several sources, such as network operators' radio access networks, internet service providers, and core networks; user data consists of details about the user's profile and location. The success or failure of mobile network services is heavily dependent on how well network operators can interpret this data [11]. When it comes to optimizing networks, effective data analysis tools are absolutely necessary. There are a number of obstacles to improving service quality with HPC Data Clusters with In-Memory Blockchain, despite the many benefits and uses for such a system.

Recent research on blockchain technology has shifted its focus to various systemic perspectives. For instance, [12] introduced an innovative design that employs network-coded distributed storage to address the issue of retention bloating in blockchains. ef. [13] investigates methods to safeguard blockchain networks against attacks from quantum computing. To enhance the hardware-level reliability of blockchain topologies, [14] provides a comprehensive recommendation. Data provenance via blockchain has been investigated in a few studies [15–17], although these have been narrowing in scope and typically dealt with very small datasets. Gabriel and Markus looked at how blockchain DLT relate to PROV standards for data provenance in their article. They looked at how DLT might facilitate data provenance in cloud-based HPC data clusters using in-memory blockchain.

Numerous novel suggestions are currently being considered in blockchain research, which is investigating a wide range of system viewpoints. To thwart Sybil and targeted assaults, Algorand [18] presents a new approach. To improve throughput, Bitcoin-NG [19] chooses a leader from each epoch to publish many blocks. Monoxide [20] lessens the load on overloaded nodes by distributing computing, storage, and memory resources optimally across various zones. The goal of sharding protocols [21,22] is to make distributed ledgers larger, whereas the goal of Hawk [23] is to make public blockchain transactions private. If you are looking for a PoS protocol that outperforms PoW blockchains in terms of efficiency and security, go no further than the Ouroboros protocol [24]. Modern publications [25–27] offer methods to enhance Byzantine Fault Tolerance (BFT). Furthermore, new consensus protocols are being developed at a rapid pace, with designs tailored to in-memory architecture or the establishment of unique security identities for participants in networks. Examples of these protocols include PoR [28] and proof-of-vote (PoV) [29].

Inkchain [30] is an additional permissioned blockchain solution that draws inspiration from Hyperledger [31], providing flexibility and improvement in various situations. Drawing inspiration from Hyperledger, BigchainDB [32] uses Practical Byzantine Fault Tolerance (PBFT) ideas to improve reliability and fault tolerance. The decentralized and immutable nature of blockchain technology is combined with the low latency, high transaction rates, and structured data indexing and querying capabilities of databases. Separately, in an effort to strengthen data integrity in distributed database systems, a two-layer blockchain architecture (2LBC) is presented in Ref. [33]. A combination of a leader-rotation strategy and PoW methods accomplishes this. One thing to note is that when it comes to bridging the gap between blockchain technology and HPC, none of the studies that have been mentioned thus far tackle the fundamental platform architecture beyond shared-nothing clusters (Figure 7.3).

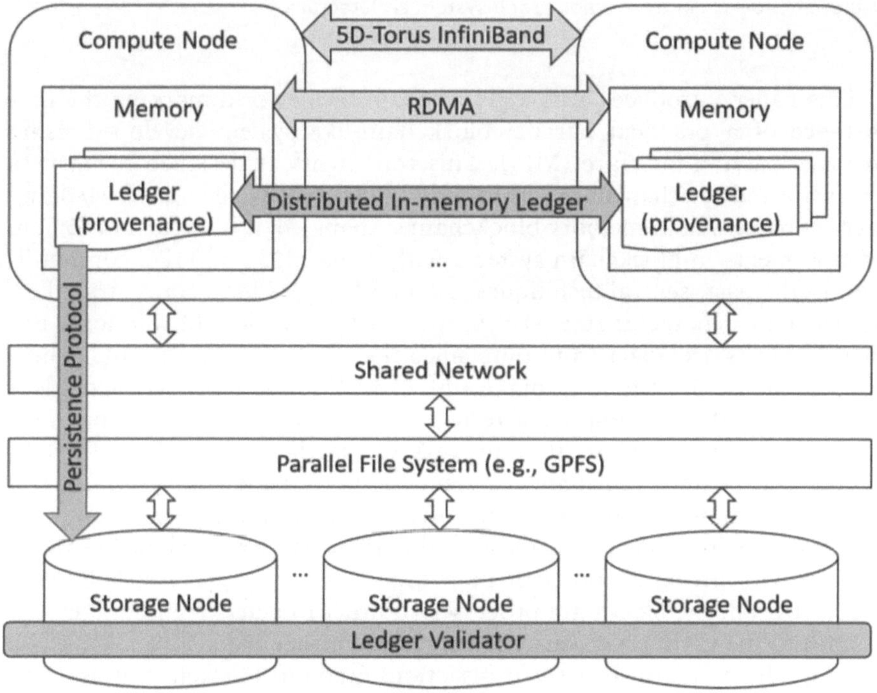

Figure 7.3 In-memory blockchain architecture.

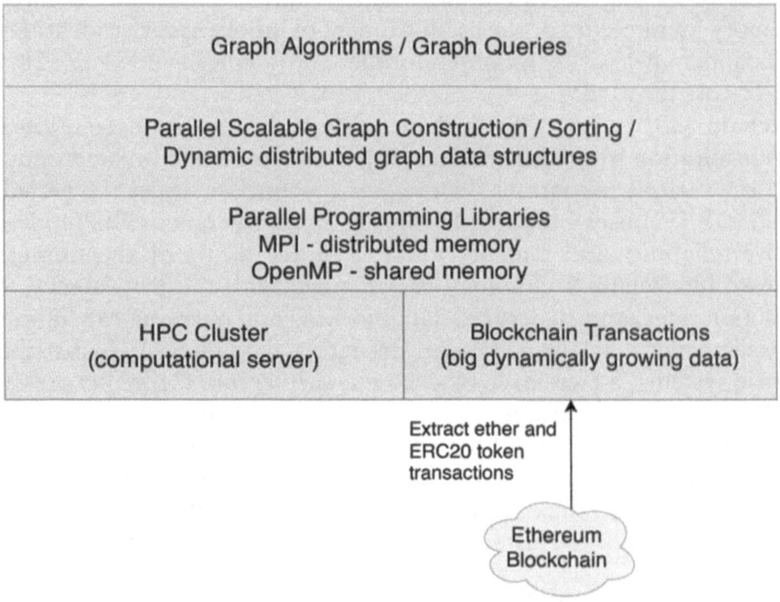

Figure 7.4 Blockchain transaction graph system architecture.

This paper introduces a novel blockchain framework, marking the first instance of a practical parallel blockchain-like system developed using message passing interface (MPI). This framework is designed to harness decentralized mechanisms within HPC systems. While acknowledging recent progress in in-memory blockchain systems [3], it is important to note advancements in blockchain systems within the MPI and HPC communities. In the past, several techniques [19,30,34,35,36] have concentrated on improving or characterizing MPI properties for a variety of solutions. But these projects are apart from ours since they do not want to build a new blockchain architecture that makes use of MPI to manage distributed ledgers in a parallel fashion. As a result, we can incorporate these previous efforts into our framework to improve MPI-specific packages even further.

Our blockchain graph analysis system's design is shown in Figure 7.4. You can get block data through a blockchain node or an Ethereum gateway, such as Infura or Cloudflare. We chose the Cloudflare gateway for block retrieval since synchronizing a node can be a time-consuming process. Following their retrieval, the blocks are processed in order to detect transactions that pertain to the ether cryptocurrency and the transfer of ERC20 tokens. We next use these files to store the transactions, and our blockchain graph system uses them as input. The Zenodo website provides access to the dataset and all of its associated formats [14]. In addition to ether, the system also retrieves 31 ERC20 token contracts from blocks. These contracts include stable coins like USDT, PAX, EURS, BUSD, GUSD, TRYB, and XAUT.

7.2.1 Applications of blockchain

Blockchain, renowned for its safety and security attributes, finds extensive applications across various domains. Its versatile applications include cryptocurrency, healthcare, supply chains, smart contracts, advertising, financial services, IoT (Internet of Things), asset management, music, financial markets, voting, banking, cybersecurity, copyright and royalties, government, facilitating international payments, weapons tracking, cost reductions, digital currency, financial management, identity management, and land registration.

7.2.2 Overview of clustering algorithms

In the contemporary era, the clustering technique in data analysis is widely employed to address the emerging challenges associated with big data. This analytical approach involves partitioning a dataset into two subsets, where one subset comprises similar instances and the other encompasses dissimilar instances [3]. Various clustering methods are utilized for this partitioning process, including bi-clustering, density-based, graph-based, grid-based, hard clustering, hierarchical, model-based, partitioning, and soft clustering. The primary objective of clustering is to group data points into clusters based on their similarities, distinguishing one cluster from another. This process involves organizing similar points into one cluster and segregating other data points into separate groups. The sequential stages of clustering are visually represented in Figure 7.5.

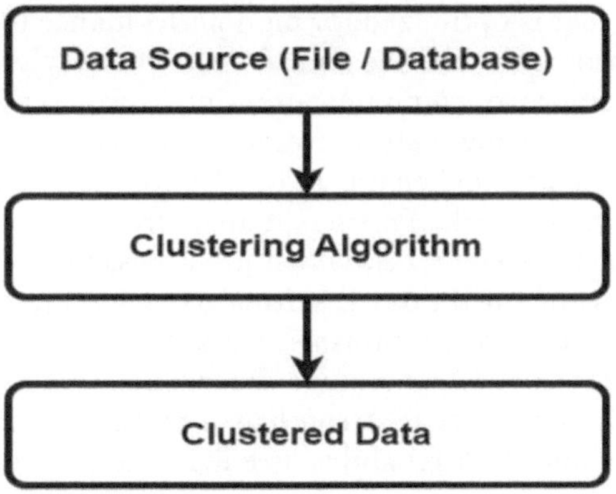

Figure 7.5 Clustering stages.

The primary goal of clustering is to recognize both similar and dissimilar characteristics or patterns within the provided data. Similar points are determined using similarity functions. Subsequent to the clustering process, class labels are assigned to the clusters, a step referred to as classification. The clustering process takes input from various data sources and yields the output as clustered data. Clustering is widely applied in diverse applications such as pattern recognition, image processing and analysis, document categorization, stock market segmentation, exploratory data analysis, World Wide Web, metrology, healthcare, and social network analysis.

7.3 THE NEW BLOCKCHAIN ARCHITECTURE FOR HPC

A typical HPC system is depicted in Figure 7.3 as a whole, together with our proposed distributed in-memory ledger. While we have shown compute nodes without discs, certain bespoke HPC systems may really have node-local discs, so keep that in mind. As an example, the burst buffer is a local Solid State Drive (SSD) storage system used by the top-ten supercomputer Cori [33] at Lawrence Berkeley National Laboratory. Since burst buffers are more suited to short-term data caching than long-term archival, they are not a good fit for ledger applications. In line with the non-time-sharing character of scientific applications, they are usually purged after job execution for performance and security concerns. We suggest a secondary ledger and validator on remote storage to address the transient nature of ledger persistence on local storage. Our proposed system's three main components—a distributed in-memory ledger, persistence protocols, and a distant persistent ledger—are depicted in the image. What follows is a discussion of each module in further depth.

At its core, the first module is about deploying a distributed ledger across computing nodes that is optimized for high-performance interconnects like InfiniBand and protocols like Remote Direct Memory Access (RDMA). The goal is to improve communication-intensive consensus methods, namely PBFT [34], by using high-performance hardware. Hyperledger [13] and other permissioned blockchains use PBFT to ensure that only verified users can access the network. This sets it apart from permissionless blockchain systems like Ethereum [14], which are accessible to everybody, and is especially important in the HPC setting for scientific applications due to the strict authentication mechanisms that are in place. Diskless computing nodes and distant persistent storage synchronization is the focus of the second module. Data validity is strengthened by the persistent ledger kept in remote storage, and data reliability is enhanced as a result of this synchronization. The main obstacle is figuring out how to export the ledger from memory to the remote parallel file system. Persisting after each transaction is not practicable because it would cause a performance bottleneck due to the considerable I/O overhead. In the third and final module, we validate the distributed ledger on persistent storage. Since every compute node updates the ledger in volatile memory (or temporary persistent local storage that is erased after a job is finished), there must be an uncompromised ground truth on permanent persistent storage to validate ledger copies in case of disasters (like when half of the compute nodes crash and lose their ledgers). Keep in mind that compute nodes that rely just on memory are not necessarily less dependable than those that have permanent storage. However, if the process that was responsible for initiating the memory were to be killed, the data that was saved in memory would be lost.

7.4 SYSTEM IMPLEMENTATION

The prototype system [21] for the blockchain architecture and consensus methods that are being suggested will be implemented using Java. You can find the project webpage at https://expolab.org/http://cse.unr.edu/hpdic/proj/imb, where you can also get the source code and other information. The prototype's main modules are now available for download, and we plan to release supplementary components and plug-ins as soon as they pass our quality assurance tests. There are more than 1,800 lines of code in the current codebase. Local ledgers are kept as independent files in the prototype's virtualized environment, and network latency is regulated by a time delay that is parameterized by random statistical distributions (with average, variation, and seed). In order to deploy the prototype on production systems, we are currently working on packaging it into Docker containers.

Our present emphasis is on the ledger and consensus, in accordance with the suggested system decomposition in Ref. [7], which specifies four primary components (ledger, consensus, cryptography, and smart contract). We use the SHA-256 hash algorithm [35] for the chained blocks. Application wrapping using pseudo transactions with dummy numerical values allows the prototype to accept multiple applications, as the current implementation does not support smart contracts. The ledger and consensus implementation are covered in detail in the following sections. The suggested PoR consensus relies on a consensus protocol between compute nodes, which is similar to traditional PoW but simplifies the computationally difficult problem found in Bitcoin [12]. Our system's PoW consists of three primary components. First, a node stores freshly made transactions, which are then sent to other nodes and shared storage, until they form a block that can be mined. Second, every node in the network that can process data, including those with shared storage, will try to verify the block and add it to their local blockchain. In mining, the node that finds the solution adds the block to its local in-memory blockchain and gets the reward. If the block does not pass validation during this round, it will be queued up on the corresponding compute node for processing at a later time. This phase is in line with the standard PoW between compute nodes, but with a modified PoW (i.e., PoR) designed for the provenance of scientific data, which is then double-validated by the storage node's ground truth. In the end, consensus is achieved when the block is added to the local blockchain by a node and stored in shared storage. After that, it is sent to other peers in the network for validation. Upon successful validation, the first node adds the block to their local blockchain and sends it out to the rest of the network. Then, when each node confirms the block's legitimacy, they update their local blockchains.

7.5 RESULTS AND DISCUSSION

This research uses Intel Core-i7 4.2 GHz CPUs and 32 GB of 2,400 MHz DDR4 memory [21]. The latency of an InfiniBand network is 1 μs [36], whereas that of an Ethernet network is 250 μs. Two other blockchain systems and two other provenance systems will be compared to the proposed in-memory blockchain prototype as part of the evaluation. Conventional Blockchain, installed on a shared-nothing cluster linked by Ethernet, is the initial blockchain system. The second system is a hypothetical blockchain that does not use permanent storage but instead makes use of high-performance networking connectivity like InfiniBand or RDMA. The lack of data durability makes it an impractical solution, but it does set the performance upper bound for the in-memory blockchain that is being proposed. Java is used to implement all three blockchain systems, and adequate optimization efforts have been made. Both the database and file systems under consideration have provenance systems as built-in modules.

Database provenance monitoring and querying is the main focus of SPADE [4], a graph database system. Built on top of the distributed file system FusionFS [38], FusionProv [1] is the provenance module for file-system provenance. The standards are structured like a bank transfer in terms of the format of the transactions. The system checks the submitted transaction's legitimacy at the beginning of each transaction. If the request is legitimate, two nodes will have their statuses (balances) modified and the new information will be sent to every other node in the network. Our investigations show that each block can include anywhere from one to ten transactions, with an average of four transactions per block.

7.5.1 Comparison to filesystems and databases

In this research, we present an in-memory blockchain as a provenance tool. Two other tools, a distributed filesystem for HPC [38] and a relational database for graph processing [4], each impose their own latency overhead. With an in-memory blockchain, the latency is unique to the node that wins the race and adds the block at the end. Unlike the other two systems, this one does not experience latency on any of its nodes. We disperse the in-memory

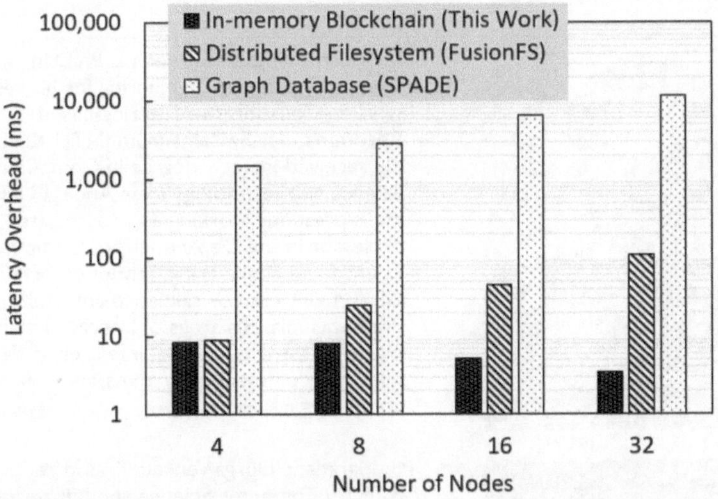

Figure 7.6 **Latency overhead of in-memory blockchain, distributed file system, and graph database.**

blockchain delay from one node to all nodes to make sure it is fair to compare. Since the database-based provenance cannot scale beyond tens of nodes, the comparison is restricted to no more than 32 nodes. At lower scales (four nodes), the in-memory blockchain has latency overhead similar to the distributed filesystem, as shown in Figure 7.6. However, when the node count increases to eight or more, the in-memory blockchain outperforms the distributed filesystem. At every size, the in-memory blockchain far surpasses the graph database. The suggested in-memory blockchain outperforms the filesystem by a factor of 32 and the database by a factor of four at the 32-node size. The fact that the amortized per-node latency drops as the number of nodes increases is another notable feature of the in-memory blockchain.

7.6 CONCLUSION

This chapter presents a new architecture for blockchain systems that primarily uses memory to store ledgers, along with a consensus process that is specifically intended for this architecture. These developments, taken as a whole, make it possible to run a blockchain-like ledger service efficiently, which guarantees trustworthy data provenance on HPC systems. Experimental validation and theoretical support both back up the proposed consensus. An evaluation of a lightweight system prototype with more than one million transactions showed a 32× acceleration when compared to provenance services based on filesystems and a four orders of magnitude acceleration when compared to provenance services based on databases.

AUTHOR DETAILS

Dr. Srikanth Thota received his Ph.D in Computer Science Engineering for his research work in Collaborative Filtering-based Recommender Systems from JNTU, Kakinada. He received M.Tech. degree in Computer Science and Technology from Andhra University. He is presently working as an associate professor in the Department of Computer Science and Engineering, School of Technology, GITAM University, Visakhapatnam, Andhra Pradesh, India. His areas of interest include machine learning, artificial intelligence, data mining, recommender systems, and soft computing.

Mr. Maradana Durga Venkata Prasad received his B.Tech. (Computer Science and Information Technology) in 2008 from JNTU, Hyderabad, and M.Tech (Software Engineering) in 2010 from Jawaharlal Nehru Technological University, Kakinada. He is a research scholar with Regd No: 1260316406 in the Department of Computer Science and Engineering, Gandhi Institute Of Technology And Management (GITAM), Visakhapatnam, Andhra Pradesh, India. His research interests include clustering in data mining, Big Data analytics, and artificial intelligence. He is currently working as an Assistant Professor in the Department of Computer Science Engineering, CMR Institute of Technology, Ranga Reddy, India.

REFERENCES

1. A. Gehani and D. Tariq, "SPADE: Support for provenance auditing in distributed environments," In *Proceedings of the 13th International Middleware Conference (Middleware)*, Montreal, QC, Canada 2012.
2. T. Clark, P. N. Ciccarese, and C. A. Goble, "Micropublications: A semantic model for claims, evidence, arguments and annotations in biomedical communications," *Journal of Biomedical Semantics*, vol. 5, no. 1, p. 28, 2014.
3. E. Pettersen, T. Goddard, C. Huang, G. Couch, D. Greenblatt, E. Meng, and T. Ferrin, "UCSF chimera: A visualization system for exploratory research and analysis," *Journal of Computational Chemistry*, vol. 25, no. 13, pp. 1605–1612, 2004.
4. R. Ranjan, D. Pandey, A. K. Rai, D. Gupta, P. Singh, P. R. Kumar, and S. N. Mohanty, "A manifold-level hybrid deep learning approach for sentiment classification using an autoregressive model," *Applied Sciences*, vol. 13, no. 5, p. 3091, 2023.

5. D. Dai, Y. Chen, P. Carns, J. Jenkins, and R. Ross, "Lightweight provenance service for high-performance computing," In *International Conference on Parallel Architectures and Compilation Techniques,* Portland, OR, USA 2017.
6. B. Shilpa, P. R. Kumar, and R. K. Jha, "LoRa DL: A deep learning model for enhancing the data transmission over LoRa using autoencoder," *The Journal of Supercomputing*, vol. 79, pp. 17079–17097, 2023.
7. X. Liang, S. Shetty, D. Tosh, C. Kamhoua, K. Kwiat, and L. Njilla, "Provchain: A blockchain-based data provenance architecture in cloud environment with enhanced privacy and availability," In *IEEE/ACM International Symposium on Cluster, Cloud and Grid Computing (CCGRID),* Madrid, Spain 2017.
8. Aravind Ramachandran and M. Kantarcioglu, "Smartprovenance: A distributed, blockchain based data provenance system," In *Proceedings of the Eighth ACM Conference on Data and Application Security and Privacy, Series, CODASPY'18,* NY, United States pp. 35–42, 2018.
9. X. Chen, S. Lin, and N. Yu, "Bitcoin blockchain compression algorithm for blank node synchronization," In *Proceedings of 11th International Conference on Wireless Communications and Signal Processing (WCSP),* Xi'an, China, pp. 1–6, October 2019.
10. R. Ranjan, D. Pandey, A. K. Rai, D. Gupta, P. Singh, P. R. Kumar, and S. N. Mohanty, "A manifold-level hybrid deep learning approach for sentiment classification using an autoregressive model," *Applied Sciences*, vol. 13, no. 5, p. 3091, 2023.
11. Z. Guo, Z. Gao, H. Mei, M. Zhao, and J. Yang, "Design and optimization for storage mechanism of the public blockchain based on redundant residual number system," *IEEE Access*, vol. 7, pp. 98546–98554, 2019.
12. Y. Xu, "Section-blockchain: A storage reduced blockchain protocol, the foundation of an autotrophic decentralized storage architecture," In *Proceedings of 23rd International Conference on Engineering of Complex Computer Systems (ICECCS 2024),* Melbourne, VIC, Australia pp. 115–125, December 2018.
13. N. Arivazhagan, K. Somasundaram, G. B. Mohammad, P. R. Kumar, et al., "Cloud-Internet of Health Things (IOHT) task scheduling using hybrid moth flame optimization with deep neural network algorithm for e healthcare systems," *Scientific Programming*, vol. 2022, pp. 1–12, 2022.
14. T. Liu, J. Wu, J. Li, and J. Li, "Secure and balanced scheme for nonlocal data storage in blockchain network," In *Proceedings of 2019 IEEE 21st International Conference on High Performance Computing and Communications; IEEE 17th International Conference on Smart City; IEEE 5th International Conference on Data Science and Systems (HPCC/SmartCity/DSS),* Zhangjiajie, China, pp. 2424–2427, 2019.
15. P. R. Kumar, G. B. Mohammad, and P. Dileep, "Real-time heart rate monitoring system using least square method," *Annals of the Romanian Society for Cell Biology*, vol. 25, no. 6, pp. 16302–16308, 2021.
16. L. Aniello, R. Baldoni, E. Gaetani, F. Lombardi, A. Margheri, and V. Sassone, "A prototype evaluation of a tamper-resistant high performance blockchain-based transaction log for a distributed database," In *13th European Dependable Computing Conference (EDCC),* Geneva, Switzerland 2017.
17. D. Dai, Y. Chen, P. Carns, J. Jenkins, and R. Ross, "Lightweight provenance service for high-performance computing," In *International Conference on Parallel Architectures and Compilation Techniques (PACT),* Portland, OR, USA 2017.

18. D. Dai, Y. Chen, D. Kimpe, and R. Ross, "Provenance-based object storage prediction scheme for scientific big data applications," In *IEEE International Conference on Big Data (BigData),* Washington, DC, USA 2014.
19. G. B. Mohammad, Selvarajan Shitharth, and P. R. Kumar, "Integrated machine learning model for an URL phishing detection," *International Journal of Grid and Distributed Computing,* vol. 14, no. 1, pp. 513–529, 2021.
20. X. Niu, R. Kapoor, B. Glavic, D. Gawlick, Z. H. Liu, V. Krishnaswamy, and V. Radhakrishnan, "Provenance-aware query optimization," In *IEEE 33rd International Conference on Data Engineering (ICDE),* San Diego, CA, USA 2017.
21. P. Mehta, S. Dorkenwald, D. Zhao, T. Kaftan, A. Cheung, M. Balazinska, A. Rokem, A. Connolly, J. Vanderplas, and Y. AlSayyad, "Comparative evaluation of big-data systems on scientific image analytics workloads," In *Proceedings of the 43rd International Conference on Very Large Data Bases (VLDB),* Washington, DC, USA 2017.
22. T. Li, C. Ma, J. Li, X. Zhou, K. Wang, D. Zhao, and I. Raicu, "Graph/z: A key-value store based scalable graph processing system," In *IEEE International Conference on Cluster Computing,* Chicago, IL, USA 2015.
23. P. R. Kumar and T. Ananthan, "Machine vision using LabVIEW for label inspection," *Journal of Innovation in Computer Science and Engineering (JICSE)*, vol. 9, no. 1, pp. 58–62, 2019.
24. D. Zhao, J. Yin, K. Qiao, and I. Raicu, "Virtual chunks: On supporting random accesses to scientific data in compressible storage systems," In *IEEE International Conference on Big Data,* Washington, DC, USA pp. 231–240, 2014.
25. D. Zhao, J. Yin, and I. Raicu, "Improving the I/O throughput for data intensive scientific applications with efficient compression mechanisms," In *International Conference for High Performance Computing, Networking, Storage and Analysis (SC'13), Poster Session,* Geneva, Switzerland 2013.
26. P. R. Kumar, "Wireless mobile charger using inductive coupling," *Journal of Emerging Technologies and Innovative Research (JETIR)*, vol. 5, no. 10, pp. 40–44, 2018.
27. H. Qin, S. Zawad, Y. Zhou, L. Yang, D. Zhao, and F. Yan, "Swift machine learning model serving scheduling: a region based reinforcement learning approach," In *Proceedings of the International Conference for High Performance Computing, Networking, Storage and Analysis (SC)*, Montreal, QC, Canada 2019.
28. E. Saillard, P. Carribault, and D. Barthou, "Static/dynamic validation of MPI collective communications in multi-threaded context," In *Proceedings of the 20th ACM SIGPLAN Symposium on Principles and Practice of Parallel Programming, PPoPP 2015. ACM,* New York, United States. pp. 279–280, 2015.
29. B. Shilpa, P. R. Kumar, and R. K. Jha, "Spreading factor optimization for interference mitigation in dense indoor LoRa networks," *IEEE IAS Global Conference on Emerging Technologies (GlobConET),* London, UK pp. 1–5, 2023.
30. C. Shou, D. Zhao, T. Malik, and I. Raicu, "Towards a provenance- aware distributed filesystem," In *TaPP Workshop, USENIX Symposium on Networked Systems Design and Implementation (NSDI),* Chicago, USA, 2013.

31. J. Sousa, A. Bessani, and M. Vukolic, "A byzantine fault-tolerant ordering service for the hyperledger fabric blockchain platform," In *48th Annual IEEE/IFIP International Conference on Dependable Systems and Networks (DSN),* Luxembourg, Luxembourg 2018.
32. J. Wang and H. Wang, "Monoxide: Scale out blockchain with asynchronized consensus zones," In *16th USENIX Symposium on Networked Systems Design and Implementation (NSDI 19),* Boston, MA, 2019. USENIX Association.
33. P. R. Kumar, "Position control of a stepper motor using LabVIEW," *3rd International Conference on Recent Trends in Electronics, Information Communication Technology (RTEICT),* Bangalore, India pp. 1551–1554, May 2018.
34. M. Zamani, M. Movahedi, and M. Raykova, "Rapid chain: Scaling blockchain via full sharding," In *Proceedings of the 2018 ACM SIGSAC Conference on Computer and Communications Security.* ACM, New York, United States pp. 931–948, 2018.
35. K. Zhang and H.-A. Jacobsen, "Towards dependable, scalable, and pervasive distributed ledgers with blockchains," In *38th IEEE International Conference on Distributed Computing Systems (ICDCS),* Vienna, Austria 2018.
36. P. R. Kumar and B. Shilpa, "An IoT-based smart healthcare system with edge intelligence computing," In S. Satpathy, S. N. Mohanty, and S. Potluri (eds.), *Reconnoitering the Landscape of Edge Intelligence in Healthcare.* CRC Press, Boca Raton, FL, pp. 31–46, 2024.

Chapter 8

DeepShield

A deep learning approach for robust fraud detection in credit financial transactions

Mulagundla Sridevi, Gouthami Velakanti, B. Deevena Raju, and Sadda Bharath Reddy

8.1 INTRODUCTION

Financial statements are prepared by accounting and finance departments and then examined by regulatory agencies like SEBI and RBI to make sure they are authentic. A nation's economic prosperity is dependent on the stability of its businesses and the security of its investors' money. The financial situation, investments, obligations, interest paid, and interest earned can all be found in the financial statements. A company's financial health is a true reflection of how its assets and liabilities have grown over time. These declarations are used by companies to get additional loans or investments. Rating agencies use them to assign credit scores, and investors use them to make educated investment decisions. Governments or third parties utilize the statements to recognize exceptional performance with prizes, while creditors use them to approve or recoup debts.

It is possible for corporations to falsify financial statements in order to increase their attractiveness to potential investors, lenders, or award givers, because these statements are used by outside parties to evaluate a company's financial health. It is crucial to confirm the legitimacy of financial statements before to making judgments, as fraud is becoming more common in emerging economies. An option for confirmation is to use a machine learning (ML) model, which can be a cataloging model trained on tagged financial records. Credible or dishonest are some of the terms used by professional auditors to classify statements.

These ML models play a crucial role in categorizing financial information within the context of digital transitions within enterprises. Not only do they reveal whether fraud is present, but they also provide an explanation for why some assertions are deemed false. When conducting manual audits, it is critical for auditors to have a firm grasp on the main causes of fraud. Although the steps for creating a classification model or extracting key components are same, it is common practice to use an additional method, such SHAP, to determine which factors are crucial. Several algorithms have

DOI: 10.1201/9781032663005-8

been investigated in the past for the purpose of detecting financial statement fraud. Classification methods like as logistic regression (LR), xgboost, adaboost, and neural networks (NNs) are commonly employed; the models are selected according to the dataset's characteristics and the level of accuracy needed. But not every algorithm can be explained. Black box models, such as NNs, xgboost, or adaboost, produce reliable classifications but do not reveal what criteria were used to arrive at that designation. So, 'black box' is a common name for these types of models.

Successful alternative methods for accurate categorization include support vectors [4,5,6] and Zipf's law [7]. NNs and belief networks [8], NNs using the backpropagation algorithm [9], NNs for risk assessment [10], preprocessing methods for data preparation in training fraud detection models [11], the importance of digit distribution [12], and other significant methodologies [13] have all been extensively studied in previous research. By employing algorithms such as regression and tree-based approaches, the authors tested 38 models with various dataset combinations in order to determine the best model and important factors [14]. The approach included training the models using different techniques and datasets that were either over-sampled or under-sampled.

Prediction accuracy in ML models was dependent on features and feature engineering, in contrast to many older models that used human feature derivation inputs into a classification layer to establish probabilities. You can reduce your reliance on feature engineering by using deep learning models. Recent studies have looked into employing deep learning algorithms like convolutional neural networks (CNNs) and DNNs to identify fraudulent activities and extract important features. CNNs were developed in Ref. [16], while DNNs have evolved over time thanks to the contributions of several scholars [15]. Although CNNs first gained traction in the image processing industry, they have now found use in other domains, such as Natural Language Processing (NLP) and fraud detection [17].

8.2 RELATED WORK

The ever-increasing volume of online purchases has put credit card fraud detection at the forefront of research priorities. Class imbalance and data's inherent volatility are two major obstacles to effective fraud detection [18]. Resampling is one of several approaches that have been developed to solve the long-standing issue of class imbalance in credit card fraud detection [19]. To do this, a training dataset must be balanced, which means either under-sampling the majority class or over-sampling the minority class [20]. Bagging, boosting, and stacking are ensemble approaches that have also been used to address class imbalance [23]. One alternative is cost-sensitive learning, which uses a cost-based approach to classifying misclassification errors and usually gives a higher cost to minority classes [24].

In addition to imbalance in number, the spatial distribution of cases from different classes greatly affects the results of classifiers. Cases close to the border between majority and minority classes, for example, are critical for correct identification, which is why techniques like Gaussian mixture under-sampling have been proposed [25]. Adapting user transaction habits leads to variations in consumption seasonality and fraud patterns, which in turn creates the difficulty of data dynamic change in transaction data. Conventional approaches to fraud detection, like CNNs, random forests (RFs), and support vector machines (SVMs), frequently presume homogeneous classes and constant data distribution. A thorough analysis of techniques used to identify credit card fraud from 1990 to 2017 reveals an emphasis on improving classifier performance through the use of combined class imbalance processing methods, with little regard for the dynamic change in data [23]. Concept drift is one way that data is always changing; research into this topic has focused on finding new ideas quickly and adapting classifier updates to account for them [5]. Because fraudulent and legitimate transactions are fundamentally different, it is critical to derive accurate representations that consistently differentiate between the two, even as fraud tactics change. Building a reliable model for detecting fraud using deep representation learning techniques is the main focus of this article [12].

8.2.1 Deep supervised representation learning

In order to improve the performance of classifiers or predictors, representation learning entails learning a new representation for the provided data, with the goal of capturing more valuable information [12]. This method has been quite effective in many fields, but it has especially shined in supervised learning-based large-scale visual categorization for feature extraction [25]. The visual domain has been the site of a great deal of research into deep representation learning, with many studies taking advantage of open-source datasets such as ImageNet, LFW, and COCO [2,7,8]. The performance limits of the related representation model are heavily influenced by the architecture of a deep NN. CNNs' depth and breadth can be tuned to control their capacity, and several designs have found success with this approach, such as ResNet, DenseNet, and BagNet [1,3,9].

To improve classification performance and create a balanced dataset, a thorough mechanism was proposed in the context of credit card fraud detection that uses K-means clustering and the genetic algorithm to generate new data samples for minority clusters [14]. Using a genetic algorithm influenced by natural selection and genetics, this method forms clusters of similar data points using unsupervised learning. Then, fresh samples for minority classes are generated. Making training sets for detecting card fraud that are more evenly distributed and with fewer classification errors is the goal. Ensemble learning was the subject of an alternative study that sought to identify instances of credit card fraud [15]. In order to make better predictions, ensemble learning uses a combination of different ML classifiers. For accurate fraud and non-fraud case identification, RFs and Artificial Neural Networks (ANNs) are employed. Using a mix of three feed-forward NNs with distinct hyperparameters and two RF classifiers with different decision trees (DT), the study acknowledges the significant monetary cost of misclassifying both legitimate and invalid transactions. Finally, the output is determined by averaging the results of the five models.

Credit card fraud detection was the subject of a comparative study in which a CNN, a multilayer perceptron layer (MPL), and a basic NN were evaluated. Using common variables in financial institution databases and conventional predictors for predictive modeling, the self-generated dataset, which included 60,000 transactions and 12 features, was constructed. The study used a learning rate of 0.001 with the 'softmax' activation function, and the imbalanced dataset was balanced by under-sampling. The accuracy levels displayed by MPL and CNN were 87.88% and 82.86%, respectively. Credit card fraud detection using a real-time deep learning model incorporating auto-encoders was presented in another work [17]. The confusion matrix, recall, accuracy, and precision were performance measurements. The majority of fraudulent transactions were predicted by non-linear auto-regression, but many genuine transactions were misclassified as well. When it came to valid transactions, LR had the best misclassification error rate, but when it came to fraud, it was very inaccurate. Here, the deep NN Auto Encoder showed consistent performance, with a better prediction rate and less misclassification error.

Credit card fraud detection using CNNs was discussed in Ref. [18] because of CNNs' capacity to find hidden fraud tendencies and reduce overfitting. Trading entropy was a new feature that was introduced as part of feature engineering, which also included creating aggregated features from transaction data. In order to ensure that the dataset was balanced, synthetic fraud samples were generated from actual fraudulent data using cost-based sampling. The assessment criteria was the F1 score, and the CNN used was six layers strong, similar to LeNet. For various sample sets, CNN outperformed NN, SVM, and RF; this was particularly true when the trading entropy feature was included. Also, Ref. [19] looked into credit card fraud detection down to the transaction level, highlighting how crucial it is to account for the passage of time in a series of transactions to account for the dynamic nature of fraud. To boost classification accuracy, statistical features were included based on actual features.

Credit card fraud detection using CNN, stacked long short-term memory (SLSTM), and a CNN-LSTM hybrid model was discussed in Ref. [20]. CNN was great at learning from very brief sequences, whereas LSTM was great at learning from very long ones. The study used principal component analysis (PCA) to reduce dimensionality using a dataset from an Indonesian bank with different non-fraud to fraud ratios. Raising the ratio enhanced the classifier's accuracy, according to the results. The order of training accuracy was as follows: SLSTM, CNN-LSTM, and CNN. With that said, taking into account the datasets' inherent imbalance, CNN outperformed CNN-LSTM and SLSTM according to area under the curve (AUC) values, highlighting the predominance of short-term linkages in fraud transaction patterns.

8.3 METHODOLOGY

The lack of publicly available datasets is a challenge for ML approaches used to detect credit card fraud. This is mostly due to the sensitive nature of financial data and the requirement to preserve user privacy. The results of ML models might change substantially across various datasets or business scenarios, and most studies in this area only use one dataset. Key research agenda items for this study include investigating performance variations across three datasets with different feature and transaction counts. Another difficulty is the class imbalance problem, which occurs when there are far fewer instances of fraud than regular transactions. One of the secondary goals of this research is to learn how different sampling strategies for dealing with class imbalance affect the performance of the models.

A lot of people have suggested using LR, SVMs, and DT to identify credit card fraud. In contrast, massive datasets might challenge these algorithms. Due to their capacity to handle enormous datasets, deep learning algorithms like CNN and LSTM are recommended for image classification and NLP, respectively. Examining the efficacy of various deep learning techniques for credit card fraud categorization is the primary goal of this research. Another important step in ML is data preprocessing. The purpose of this research is to examine the relationship between data preprocessing methods and classification performance in the context of detecting credit card fraud.

8.3.1 One-dimensional CNN (1DCNN)

In the realm of image processing in particular, the deep learning technique known as a CNN is frequently linked with geographical data. While CNNs share similarities with ANNs, they differ in the convolution layers they use, which can have different numbers of channels. These layers are used for hidden layer processing. 'Convolution' refers to the process of extracting crucial information from data by means of moving filters. One reason CNN is so popular in image processing is its ability to automatically reduce features, which makes it more resistant to overfitting. Therefore, considerable data preprocessing is not necessary for training CNN. Minimizing processing by reducing image size without losing crucial information for making predictions is the primary purpose of CNN in image processing [21]. Features maps, channels, pooling, stride, and padding are essential ideas in CNNs. Contrary to the well-known multilayer perceptron (MLP) network,

CNN does not have a fully connected topology from layer to layer. To reduce the number of parameters in a CNN model, CNN uses a constant weight parameter for each filter, in contrast to MLP where each node has variable associated weights. The feature detection process is also improved by the pooling method, making it more resistant to changes in element size and position in the image. The study uses 2DCNN and 1DCNN to categorize situations as either fraud or non-fraud. The 2DCNN is applied to the 30-feature European Card Dataset. To feed into the 2DCNN model, each transaction sample is transformed into a two-dimensional picture.

8.3.2 Long short-term memory network

Long short-term memory, often known as LSTM, is classified as a type of NN known as Recurrent Neural Networks (RNNs). Long short-term memory, often known as LSTM, is classified as a type of NN known as Recurrent Neural Networks (RNNs) is designed as memory-enabled NNs, as opposed to regular NNs, which are unable to remember past data and need to retrain for each new task. The vanishing gradient problem, however, makes short-term memory a common challenge for RNNs. Put simply, during backpropagation, the gradient decreases as it travels backward in the network, resulting in minimal modifications to the weights. This means that RNNs can only store short-term information, as the earlier layers of the network do not learn much and cannot recall early examples in long sequences. LSTM networks overcome RNNs' short-term memory problem by including a network memory (cell state) that is passed across each step. The forget gate, input gate, and output gate are the essential components that are linked to each phase. Each gate in the process has a specific function: the forget gate selects data to be retained from the previous phase, the input gate selects data to be added from the current step, and the output gate selects the concealed state for the next step (Table 8.1).

Crucial to overcome the constraints of short-term memory, the cell state remembers important information beginning with the earliest examples in the sequence. This is in contrast to the hidden state, which has limited

Table 8.1 LSTM structure

Layer 1	Input
Input shape	(1, number of features)
Layer 2	Dense
Number of LSTM blocks	50
Activation function	ReLU
Layer 3	Output
Number of nodes	1
Activation function	Sigmoid

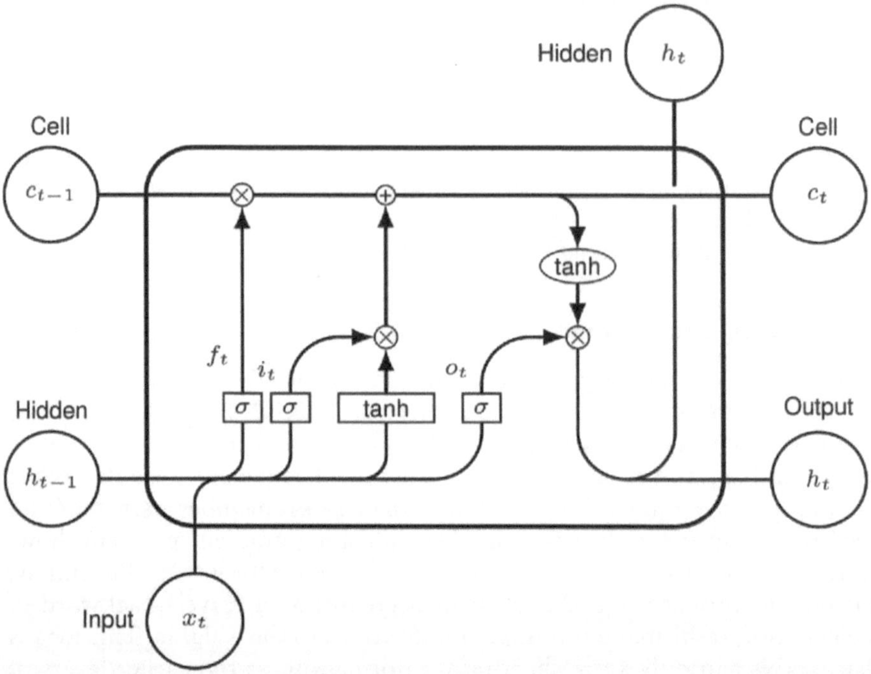

Figure 8.1 A long short-term memory cell.

long-term memory but remembers the model's past inputs. In Figure 8.1, each dot represents a single step in the full flow of an LSTM cell [22].

The forget gate is involved in the initial phase, which is seen in the red dotted box. By merging the prior hidden state (h_t−1) with the present input (x_t) and applying a 'sigmoid' activation function, we can produce an output that can be either 0 (total forget) or 1 (total retain). The next stage, indicated by the yellow dot in the box, is the input gate. Here, the 'sigmoid' and 'tanh' activation functions are applied to the past hidden state and the present input, respectively, and the resulting values are multiplied. While the 'sigmoid' function decides what data to keep from the present regulation, the 'tanh' function controls the model. Computing the cell state is the next step, as seen in the purple dotted box. The new cell state (C_t) is obtained by multiplying the previous cell state (C_t−1) with the forget gate's output and then adding the product to the input gate's output in a pointwise fashion. Calculating the new hidden state (h_t) is the responsibility of the output gate, the last step depicted in the dotted box. Before being multiplied by the output of the 'sigmoid' activation function, which takes the previous concealed state and current input as input, the new cell state goes via a 'tanh' activation function.

8.4 RESULTS AND DISCUSSION

In this study, we use datasets with different sample sizes and feature densities to evaluate classifiers' ability to detect credit card fraud. This is accomplished by making use of three separate datasets: ET Data (Europe), SCD (Small Card Data), and TCD (Tall Card Data). These datasets are very imbalanced, with many fewer occurrences of fraud than normal transactions; this is typical of credit card fraud datasets. A '0' representing no fraud and a '1' representing fraud were used to label all three datasets used in this investigation. What follows is a breakdown of each dataset along with the percentage of class imbalance and other pertinent characteristics.

8.4.1 European card data

This dataset covers two days of transaction data for European cardholders in September 2013. It was retrieved from Kaggle and sourced from the Machine Learning Group of Université Libre de Bruxelles. With 31 characteristics and 284,807 examples, it is quite extensive. Out of all the samples, just 492 have been found to be instances of fraud, which is <0.172% of the total dataset. With the exception of 'Time' and 'Amount', all characteristics in the dataset have been transformed using PCA in order to preserve client privacy and safeguard the sensitive nature of transaction details. 'Time' shows the entire length of time that has passed since the first dataset sample, in seconds, and 'length' shows the total amount of money that has changed hands. The electricity consumption dataset of Uruguay (ECD) dataset is used in this investigation.

8.4.2 Small card data

This 3,075-sample dataset with 12 characteristics was sourced from Kaggle and is on the smaller side. There are numerical features and categorical features split evenly. Out of a total of 3,075 samples, 448 (or 14.6%) were found to be instances of fraud. Dataset features include merchant ID, date of transaction, average daily transaction amount, amount declined, number of declines per day, foreign transaction status, high-risk country status, average daily chargeback amount, average 6-month chargeback amount, frequency of 6-month chargebacks, and fraudulent status. We call this dataset SCD since its rows and columns are quite small.

8.4.3 Evaluation metrics

The datasets have a large class imbalance, making accuracy the wrong criterion to use when comparing models. In credit card fraud detection systems, for example, the main objective is to catch all instances of fraud while avoiding false alarms, or legal transactions mistakenly recognized as fraudulent. The nature of the solution dictates the choice of assessment metric. This research makes use of the confusion matrix, which classifies cases as either positive (fraud) or negative (non-fraud). False positives indicate non-fraud cases anticipated as fraud, genuine negatives imply fraud instances forecast as non-fraud, and true negatives indicate accurately predicted fraud cases. Take a look at the F1 score, accuracy, precision, and recall equations down below to learn more about the assessment measures.

$$\text{Accuracy} = \frac{\text{TP} + \text{TN}}{\text{TP} + \text{FP} + \text{TN} + \text{FN}} \tag{8.1}$$

$$\text{Recall} = \frac{\text{TP}}{TP + \text{FN}} \tag{8.2}$$

$$\text{Precision} = \frac{\text{TP}}{TP + \text{FP}} \tag{8.3}$$

$$\text{F1 Score} = \frac{2 * (\text{Precision} \times \text{Recall})}{\text{Precision} + \text{Recall}} \tag{8.4}$$

Improvements in accuracy can be achieved by decreasing the occurrence of false positives, which are correlated with positive projected values. Precision is an appropriate indicator when the cost of false positives is substantial. The number of false negatives should be minimized in order to maximize recall, as it is connected to actual positives (Eq. 8.3). Achieving high recall is typically prioritized in scenarios when the cost of false negatives is large. Striking a balance between the two is of the utmost importance for detecting credit card fraud. A high recall, but poor accuracy, precision, and F1 score are the result of labeling all samples as fraudulent. However, if we assume that all samples are legitimate, we will have great accuracy but no recall, and our precision and F1 score will be unknown. To conduct comparisons, this research makes use of all four metrics: accuracy, precision, recall, and F1 score.

8.5 DISCUSSION

8.5.1 Data preprocessing

Data preprocessing is the first step of the experiment, and it entails going over all three datasets by hand and applying statistical processes. Optimal output from classifiers is the goal of data preprocessing, which entails presenting them with refined input. Missing data, categorical features, variable scalability, and high dimensionality are just a few of the variables that might affect classifier effectiveness. Data exploration and scaling are two preprocessing

Table 8.2 Data exploration

ECD	
Number of rows	284,807
Number of columns	31
Feature type	Numeric
Missing values	None
Dropped features	None
Categorical to numeric	None
Smaller sample used	No
SCD	
Number of rows	3,075
Number of columns	12
Feature type	Numeric + categorical
Missing values	3,075
Dropped features	'Transaction date'
Categorical to numeric	'Merchant_id', 'Is declined', 'isForeignTransaction', 'isHighRiskCountry', 'isFradulent'
Smaller sample used	No
TCD	
Number of rows	10,000,000
Number of columns	9
Feature type	Numeric
Missing values	None
Dropped features	'custID'
Categorical to numeric	None
Smaller sample used	Yes

approaches used in this work, together with a test-train split. Table 8.2 provides important details about the data exploration method. There were no missing values found in the ECD dataset, and all features were numerical. In order to sanitize the data, no features were removed. The 'Transaction Date' feature was eliminated from the SCD dataset because it included no values at all, and all other categorical characteristics were transformed into numerical ones. All data from the ECD and SCD sets were used. A subset of the full dataset was used for the TCD dataset. The 'custID' feature was removed from the TCD dataset since it included only unique values and did not contribute any information. All of the values in the dataset were numerical.

Next, we look at the association between features for each dataset after we have explored the data. One statistical tool for exploring the relationships between variables is the correlation coefficient, which can take on values between –1 and 1. No association is indicated by a correlation value of 0, an inverse relationship by a negative correlation, and a direct relationship by a positive correlation. In order to reduce the data's dimensions, it is helpful to identify correlations so that features with similar behavior can be eliminated. Training times and classification performance are both enhanced by using lower-dimensional data. Feature correlation in the SCD dataset is shown in Figure 8.2. The lack of a negative correlation between features and the uneven distribution of correlation across the SCD dataset are both obvious. Greater transaction quantities are associated with higher average daily transaction amounts. There is a robust relationship between the daily chargeback amount, the 6-month chargeback amount, and the 6-month chargeback frequency. Among the many factors used to categorize fraud, the 'high-risk nation' attribute stands out. No dimensionality reduction is done because the dataset is small. The performance is assessed using the test dataset, and Figure 8.3 shows a comparison of various class imbalance ratios in the ECD dataset.

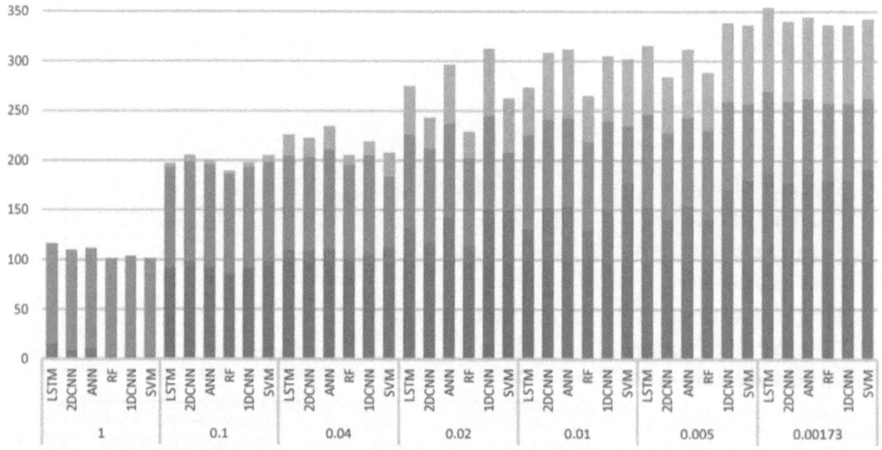

Figure 8.2 Class imbalance comparison (test data ECD).

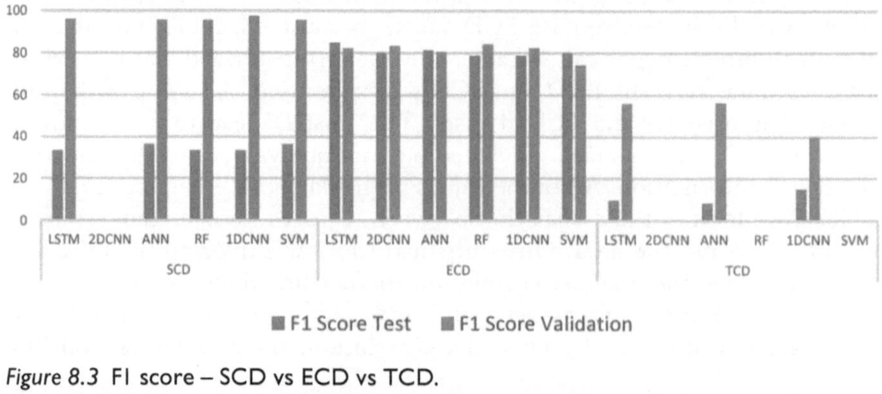

Figure 8.3 F1 score – SCD vs ECD vs TCD.

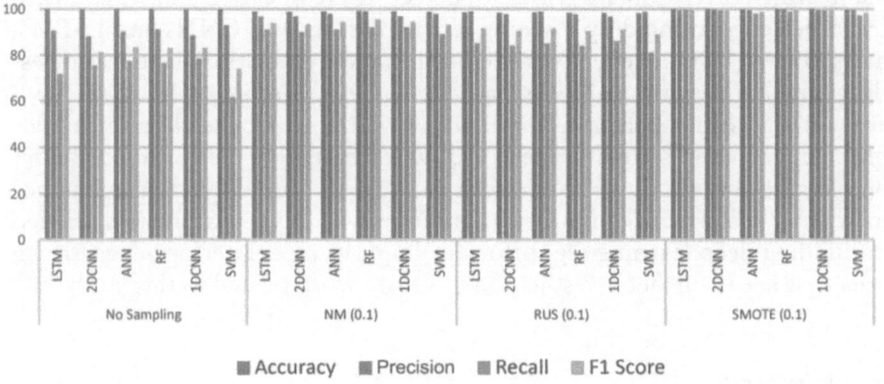

Figure 8.4 Sampling method comparison (validation data ECD).

In the beginning, recall is the highest metric while all the others are at their lowest, according to the observation. Recall drops while F1-score, accuracy, and precision go up when the class imbalance goes up. Without a 1:1 class imbalance, the model accurately predicts fraud instances but has poor accuracy when it comes to non-fraud cases. Due to the substantially higher proportion of non-fraud instances compared to fraud cases in the test data, the achieved accuracy is low. Accuracy, measured in terms of the proportion of correct predictions relative to total predictions, improves noticeably as class imbalance rises. As the disparity between classes widens, F1 score and accuracy both rise. The increasing class imbalance is the reason behind this, as it allows the model to generalize more successfully with more training cases.

The determination of which algorithm exhibits the highest level of performance was one of the key objectives of this study. Figure 8.4 shows that although traditional algorithms and deep learning approaches both performed similarly, LSTM performed somewhat better than the other algorithms. Some other things we noticed during our experiments: (i) Deep learning methods using the TensorFlow library and GPU computing reduced training time compared to traditional algorithms like SVM and RF, especially on big datasets. (ii) Misclassification increased with increasing epoch count. (iii) All Near Miss Algorithm variants gave us the same results.

8.6 CONCLUSION AND FUTURE SCOPE

The prevalence of credit card theft is on the rise, and con artists are always coming up with new techniques to steal money from banks. Because frauds are always changing, it is crucial to have a durable classifier. Reducing the number of false positives while increasing the accuracy of fraud case predictions is the main objective of fraud detection systems. The input data has a major impact on the model's performance, and the efficacy of ML approaches differs across various business cases. A large number of features, transactions, and feature correlation are critical in detecting credit card fraud. Credit card fraud detection using deep learning techniques, such as CNNs and LSTMs, has shown to be more effective than using standard algorithms. In this investigation, LSTM with 50 blocks had the highest F1-score of 84.85%, however all algorithms performed similarly. In order to fix the class imbalance problem, we used sampling methods, which made our performance better on old cases but much worse on fresh data. As the degree of class imbalance increased, performance on unseen data improved. Hyperparameters utilized in building deep learning algorithms to improve model performance are an area that needs further investigation in future work related to this study.

REFERENCES

1. R.-C. Chen, T.-S. Chen, and C.-C. Lin, "A new binary support vector system for increasing detection rate of credit card fraud," *International Journal of Pattern Recognition and Artificial Intelligence*, vol. 20, no. 2, pp. 227–239, 2006.
2. S. Xuan, G. Liu, Z. Li, L. Zheng, S. Wang, and C. Jiang, "Random forest for credit card fraud detection," In *Proceedings of IEEE 15th IEEE International Conference on Networking, Sensing and Control (ICNSC)*, pp. 1–6, 2018.
3. B. Shilpa, P. R. Kumar, and R. K. Jha, "LoRa DL: A deep learning model for enhancing the data transmission over LoRa using autoencoder," *The Journal of Supercomputing*, vol. 79, pp. 17079–17097, 2023.
4. D. Malekian and M. R. Hashemi, "An adaptive profile based fraud detection framework for handling concept drift," In *Proceedings of 10th International ISC Conference on Information Security and Cryptology (ISCISC)*, pp. 1–6, 2013.
5. R. Ranjan, D. Pandey, A. K. Rai, D. Gupta, P. Singh, P. R. Kumar, and S. N. Mohanty, "A manifold-level hybrid deep learning approach for sentiment classification using an autoregressive model," *Applied Sciences*, vol. 13, no. 5, p. 3091, 2023.
6. J. Deng, W. Dong, R. Socher, L.-J. Li, K. Li, and L. Fei-Fei, "ImageNet: A large-scale hierarchical image database," In *Proceedings of the IEEE Conference on Computer Vision and Pattern Recognition (CVPR)*, pp. 248–255, 2009.
7. G. B. Huang, M. Mattar, T. Berg, and E. Learned-Miller, "Labeled faces in the wild: A database for studying face recognition in unconstrained environments," In *Proceedings of Workshop Faces Real-Life Images, Detection, Alignment, Recognition*, pp. 1–14, 2008.
8. N. Arivazhagan, K. Somasundaram, G. B. Mohammad, P. R. Kumar, et al., "Cloud-Internet of Health Things (IOHT) task scheduling using hybrid moth flame optimization with deep neural network algorithm for e-healthcare systems," *Scientific Programming*, vol. 2022, pp. 1–12, 2022.
9. K. He, X. Zhang, S. Ren, and J. Sun, "Deep residual learning for image recognition," In *Proceedings of the IEEE Conference on Computer Vision and Pattern Recognition (CVPR)*, pp. 770–778, 2016.
10. G. Huang, Z. Liu, L. Van Der Maaten, and K. Q. Weinberger, "Densely connected convolutional networks," In *Proceedings of the IEEE Conference on Computer Vision and Pattern Recognition (CVPR)*, pp. 4700–4708, 2017.

11. P. R. Kumar, G. B. Mohammad, and P. Dileep, "Real-time heart rate monitoring system using least square method," *Annals of the Romanian Society for Cell Biology*, vol. 25, no. 6, pp. 16302–16308, 2021.
12. W. Liu, Y. Wen, Z. Yu, and M. Yang, "Large-margin softmax loss for convolutional neural networks," In *Proceedings of the International Conference on Machine Learning*, pp. 507–516, 2016.
13. Irina Sakharova, "Payment card fraud: Challenges and solutions," *2012 IEEE International Conference on Intelligence and Security Informatics*, Arlington, VA, pp. 227–234, 2012.
14. P. R. Kumar and T. Ananthan, "Machine vision using LabVIEW for label inspection," *Journal of Innovation in Computer Science and Engineering (JICSE)*, vol. 9, no. 1, pp. 58–62, 2019.
15. K. T. Hafiz, S. Aghili, and P. Zavarsky, "The use of predictive analytics technology to detect credit card fraud in Canada," *2016 11th Iberian Conference on Information Systems and Technologies (CISTI),* Las Palmas, pp. 1–6, 2016.
16. P. R. Kumar, "Wireless mobile charger using inductive coupling," *Journal of Emerging Technologies and Innovative Research (JETIR)*, vol. 5, no. 10, pp. 40–44, 2018.
17. S. Makki, Z. Assaghir, Y. Taher, R. Haque, M. Hacid, and H. Zeineddine, "An experimental study with imbalanced classification approaches for credit card fraud detection," *IEEE Access*, vol. 7, pp. 93010–93022, 2019.
18. I. Benchaji, S. Douzi, and B. E. Ouahidi, "Using genetic algorithm to improve classification of imbalanced datasets for credit card fraud detection," In *International Conference on Advanced Information Technology, Services and Systems,* pp. 220–229. Springer, Cham, 2018.
19. B. Shilpa, P. R. Kumar, and R. K. Jha, "Spreading factor optimization for interference mitigation in dense indoor LoRa networks," *IEEE IAS Global Conference on Emerging Technologies (GlobConET),* pp. 1–5, 2023.
20. I. Sadgali, N. Sael, and F. Benabbou, "Fraud detection in credit card transaction using neural networks," In *Proceedings of the 4th International Conference on Smart City Applications,* pp. 1–4. 2019.
21. Y. Abakarim, M. Lahby, and A. Attioui, "An efficient real time model for credit card fraud detection based on deep learning," In *Proceedings of the 12th International Conference on Intelligent Systems: Theories and Applications,* pp. 1–7, 2018.
22. P. R. Kumar, "Position control of a stepper motor using LabVIEW," *3rd International Conference on Recent Trends in Electronics, Information Communication Technology (RTEICT),* pp. 1551–1554, 2018.
23. B. Wiese and C. Omlin, "Credit card transactions, fraud detection, and machine learning: Modelling time with LSTM recurrent neural networks," In M. Bianchini, M. Maggini, and F. Scarselli (eds.), *Innovations in Neural Information Paradigms and Applications.* Springer, Berlin, Heidelberg, pp. 231–268, 2009.
24. Y. Heryadi and H. L. H. S. Warnars, "Learning temporal representation of transaction amount for fraudulent transaction recognition using CNN, Stacked LSTM, and CNN-LSTM," *2017 IEEE International Conference on Cybernetics and Computational Intelligence (CyberneticsCom),* Phuket, pp. 84–89, 2017.
25. P. R. Kumar and B. Shilpa, "An IoT-based smart healthcare system with edge intelligence computing," In S. Satpathy, S. N. Mohanty, and S. Potluri (eds.), *Reconnoitering the Landscape of Edge Intelligence in Healthcare.* CRC Press, Boca Raton, FL, pp. 31–46, 2024.

Chapter 9

Integrating blockchain with big data analytics for enhanced IoT security and efficiency

Sumaiya Shaikh, Saba Sheiba, and Mulagundla Sridevi

9.1 INTRODUCTION

Fraud detection is a vital process in safeguarding against the nefarious schemes of con artists seeking to unlawfully obtain money or assets through deceptive means [1]. In today's complex economic landscape, where financial crimes such as money laundering and fraud pose significant threats to the stability of the global economy, the ability to swiftly identify and mitigate fraudulent activities is paramount. Utilizing advanced technologies like artificial intelligence (AI) and machine learning (ML), organizations across various sectors such as government, banking, insurance, healthcare, and law enforcement employ sophisticated techniques to detect and prevent fraud before it wreaks havoc [2]. These efforts encompass a diverse array of fraudulent schemes, ranging from consumer fraud, intellectual property theft, and corruption to insurance and banking fraud, asset misappropriation, and more. Consumer fraud targets individuals through deceptive practices like bogus telemarketing, email scams, and Ponzi schemes, while intellectual property theft involves the illicit acquisition and trading of proprietary information. Corruption encompasses a spectrum of unethical and illegal activities such as bribery, extortion, and kickbacks, posing serious challenges to governance and transparency [3]. Within the financial realm, fraud manifests in various forms, including insurance fraud through false claims, fraudulent bankruptcies, and asset misappropriation such as skimming cash or misusing company resources. Additionally, authorized push payment and account takeover schemes exploit vulnerabilities in payment systems and financial accounts to siphon funds illicitly [4]. Other common types of fraud include phishing, identity theft, telephone or utility fraud, investment fraud, and lottery or sweepstakes scams, each posing unique threats to individuals and organizations alike. Phishing exploits networks and email platforms to deceive victims into divulging sensitive information, while identity theft involves the unauthorized access to personal data for criminal purposes [5]. Telephone or utility fraud relies on impersonation tactics to extract private information or illicit payments, while investment fraud dupes' victims into investing in fraudulent schemes or chit funds, leading to financial losses.

DOI: 10.1201/9781032663005-9

Figure 9.1 Fraud detection.

Similarly, lottery or sweepstakes fraud preys on individuals' hopes by promising non-existent prizes in exchange for upfront fees or taxes. The fight against fraud encompasses a multifaceted approach, integrating strategies to combat money laundering, cyberattacks, forged documents, and other illicit activities [6]. By employing advanced technologies and robust preventive measures, organizations can mitigate the risks posed by fraudulent actors, safeguarding the integrity of financial systems and protecting individuals and businesses from economic harm [7] (Figure 9.1).

Data warehousing serves as a crucial pillar in the realm of data management, particularly tailored to bolster and facilitate business intelligence endeavors. It consolidates diverse datasets within a centralized repository, the data warehouse, primarily tasked with facilitating queries and analyses on historical data [8]. Its fundamental objective lies in extracting pertinent insights. However, this reservoir of information also becomes a double-edged sword, as malevolent entities exploit it to devise intricate schemes beyond their own detection thresholds. The creation of association rules emerges as a pivotal strategy in navigating through the vast expanse of data [9]. By discerning frequent if-then patterns, these rules unveil significant relationships. The computation of support and confidence metrics further refines these associations, empowering organizations to unravel critical insights buried within their data reservoirs. Subsequently, leveraging IF/ELSE patterns to scrutinize the established association rules unveils underlying customer approval patterns. This analytical framework not only aids in understanding customer behaviors but also fortifies customer authentication mechanisms. Through meticulous analysis, fraudulent attempts to circumvent authentication measures can be thwarted, bolstering system security [10]. Customer authentication stands paramount in the battle against fraudulent activities, as breaches often stem from identity failures. Successful authentication ensures system integrity, whereas failures open the floodgates to potential breaches. To counteract such vulnerabilities, authentication processes must emit signals or alerts, notifying both the system and the customer of any potential compromise [11]. In the event of system failures or authentication breaches, proactive measures come into play. The system must promptly generate alerts, signaling unauthorized or anomalous activities. These alerts serve as early warnings, enabling swift intervention to mitigate potential threats and safeguard organizational assets [12]. In essence, the synergy between data warehousing, association rule creation, IF/ELSE pattern analysis, customer authentication, and alert mechanisms forms a robust defense against fraudulent activities [13]. By harnessing the power of data and deploying sophisticated analytical techniques, organizations can stay one step ahead of fraudsters, ensuring the integrity and security of their systems and data assets.

9.2 RELATED WORK

The integration of blockchain technology with big data analytics for bolstering Internet of Things (IoT) security has garnered significant attention in recent years. A plethora of research endeavors and scholarly works have delved into this interdisciplinary domain, aiming to explore the synergies between these cutting-edge technologies [14]. Here, we present an extensive review of the existing literature, encompassing various studies, surveys, and practical perspectives on this compelling subject matter. One notable contribution to the field comes from Doe and Smith, who conducted a comprehensive review elucidating the potential of blockchain-enabled big data analytics in fortifying IoT security [15]. Their work elucidates various use cases, challenges, and opportunities, emphasizing the pivotal role of blockchain in ensuring data integrity, confidentiality, and traceability within IoT ecosystems. Similarly, Johnson and Williams offer a survey that provides a panoramic overview of the synergies between blockchain, big data analytics, and IoT security. Their study explores the integration of blockchain's decentralized ledger for securing IoT data exchanges, complemented by the role of big data analytics in real-time threat identification [16].

The authors in Ref. [17] contribute to the literature with a review focusing on the integration of blockchain and big data analytics for securing IoT applications. Their work highlights the symbiotic relationship between these technologies, emphasizing blockchain's role in establishing trust and immutability, coupled with big data analytics' capacity for predictive security analytics. Furthermore the authors in Ref. [18] present a systematic literature review synthesizing key findings from existing research on securing IoT devices through blockchain and big data analytics integration. Their study underscores the significance of data integrity, access control, and device authentication in ensuring robust IoT security frameworks.

The authors in Ref. [19] shed light on the challenges and opportunities inherent in blockchain-based big data analytics for IoT security. Their work addresses technical, regulatory, and scalability issues, emphasizing the importance of interoperability and data privacy in designing effective solutions. The authors in Ref. [20] offer a practical perspective, providing case studies and implementation strategies for deploying blockchain-enabled IoT security solutions. Their work emphasizes best practices and real-world challenges, offering insights gleaned from practical deployments.

In Ref. [21], the authors propose a novel framework integrating blockchain-driven big data analytics for IoT security enhancement. Their work outlines architectural components and security mechanisms, emphasizing scalability and resilience to cyber threats. Additionally, the authors in Ref. [22] conduct a comparative analysis of different approaches to IoT security enhancement, evaluating the effectiveness of blockchain and big data analytics integration. Their study offers insights into consensus mechanisms, data processing techniques, and scalability considerations, aiding in the design of resilient IoT security solutions.

In Ref. [23], the authors offer a comparative analysis that sheds light on the effectiveness of different approaches to IoT security enhancement. Their study evaluates various consensus mechanisms, data processing techniques, and scalability considerations inherent in blockchain and big data analytics integration. By examining the trade-offs and performance metrics associated with each approach, they provide valuable insights into designing resilient and efficient IoT security solutions tailored to specific use cases and deployment scenarios.

Furthermore, the authors in Ref. [24] propose a novel framework that leverages blockchain-driven big data analytics to enhance IoT security. Their framework delineates architectural components, data flow mechanisms, and security protocols designed to withstand cyber threats and ensure data integrity across IoT ecosystems. By integrating blockchain's decentralized ledger with big data analytics' processing capabilities, their framework offers a scalable and resilient solution for securing IoT devices and data transmissions.

In addition to addressing technical challenges, ethical considerations and user privacy concerns are paramount in the integration of blockchain with big data analytics for IoT security enhancement. Researchers such as in Ref. [25] delve into the ethical implications of deploying blockchain-enabled IoT security solutions, emphasizing the importance of transparent governance models, consent frameworks, and data protection mechanisms. By fostering trust and accountability, ethical considerations play a crucial role in ensuring the acceptance and adoption of blockchain-driven IoT security frameworks by stakeholders and end-users alike.

Moreover, practical perspectives provided in Ref. [26] offer invaluable insights into the deployment challenges and real-world applications of blockchain-enabled big data analytics for IoT security. Through case studies and implementation strategies, they elucidate best practices for overcoming technical hurdles, integrating legacy systems, and aligning security measures with organizational objectives. By bridging the gap between theory and practice, their work facilitates the adoption and implementation of blockchain-driven IoT security solutions in diverse industry domains [27].

Finally, the exploration of integrating blockchain with big data analytics for enhanced IoT security and efficiency encompasses a broad spectrum of research endeavors, spanning theoretical frameworks, comparative analyses, practical implementations, and ethical considerations [28]. By addressing technical challenges, regulatory compliance requirements, and user privacy concerns, researchers are paving the way for innovative solutions that safeguard IoT ecosystems against emerging cyber threats while unlocking new opportunities for operational optimization and value creation.

9.3 ENHANCED IOT SECURITY AND EFFICIENCY WITH BLOCKCHAIN TECHNOLOGY

Detecting fraudulent activities within IoT ecosystems has become increasingly crucial, necessitating the integration of blockchain technology with big data analytics to enhance IoT security and efficiency. The convergence of these cutting-edge technologies offers a comprehensive approach to analyzing vast and heterogeneous datasets, enabling proactive detection of anomalies and suspicious patterns indicative of fraudulent behavior [29].

The integration of blockchain with big data analytics for enhanced IoT security and efficiency entails a systematic process aimed at leveraging the strengths of each technology. Initially, data collection involves aggregating information from various IoT devices, sensors, and systems, including user logs, device telemetry, and environmental data. This data, whether structured or unstructured, provides valuable insights into device behavior, user interactions, and network activities [30].

Data integration is crucial in consolidating disparate datasets into a unified format, enabling seamless analysis and correlation of information across different sources. Technologies such as Apache Hadoop and Apache Spark play a vital role in processing and integrating these large-scale datasets, facilitating comprehensive analysis and decision-making. Data preprocessing involves refining the dataset to address missing values, outliers, and

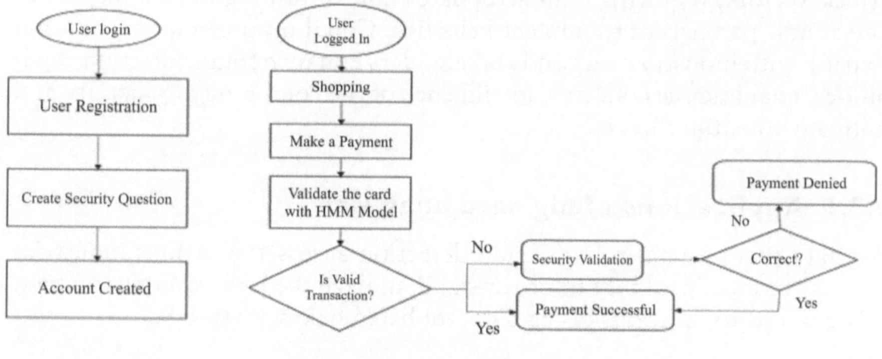

Figure 9.2 **Work flow of fraud transaction detection.**

inconsistencies, ensuring data quality and reliability. Standardization of data formats and normalization of variables are essential steps to enhance the accuracy of fraud detection algorithms. Feature engineering further enhances the dataset by extracting relevant features and creating new variables that capture meaningful insights into device behavior and network activities. ML models are employed to build predictive models for fraud detection, leveraging historical data to identify patterns and anomalies indicative of fraudulent behavior. Techniques such as neural networks, support vector machines, and decision trees are commonly used to train these models, enabling automated detection of suspicious activities in real time. Real-time processing is essential for timely detection and response to fraudulent activities, leveraging technologies such as Apache Flink and Apache Kafka for stream processing and analysis. These real-time processing frameworks enable rapid detection of anomalies and deviations from normal behavior, facilitating immediate intervention to mitigate potential threats (Figure 9.2).

Anomaly detection techniques, including statistical methods and clustering algorithms, are employed to identify unusual patterns or behaviors that may indicate fraudulent activity within IoT ecosystems. Behavioral analysis establishes baseline behavior for devices and users, enabling the detection of deviations and anomalies that may signify security breaches or fraudulent behavior. Graph analytics is utilized to analyze complex relationships and networks within IoT ecosystems, identifying suspicious connections and activities that may indicate coordinated attacks or fraudulent behavior. Scalability and performance are crucial considerations in deploying blockchain-enabled big data analytics solutions for IoT security, requiring the use of distributed computing frameworks and cloud-based solutions to handle large-scale data processing efficiently. Continuous monitoring and adaptation are essential to stay ahead of emerging threats and evolving attack vectors, requiring regular updates and retraining of ML models to detect new patterns of fraudulent behavior. Collaboration and information sharing with industry peers and stakeholders enhance fraud detection capabilities, enabling the collective intelligence of the community to identify and mitigate potential threats.

9.3.1 Applications of big data analytics

Big data plays a vital role in fraud detection across the various industries because of its high ability to process and analyze the large volumes of data. Some of the applications of big data are listed below (Figure 9.3).

9.4 FRAUD DETECTION WITH BLOCKCHAIN TECHNOLOGY

Blockchain is a decentralized ledger technology that can be used to transparently and safely record transactions. It is ideally suited for tracking financial transactions and identifying any questionable activity due to its intrinsic qualities. In today's evolving financial landscape, safeguarding against the fraud has emerged as concern for institutions striving to uphold trust and security. Traditional approaches have often fallen short due to their limitations in transparency, data integrity and operational efficiency. However, a revolutionary solution has emerged by the blockchain. With this potential to reshape anti-financial crime strategies. Some key characteristics and components of blockchain technology are as follows.

Blockchain technology is a revolutionary system characterized by several key features. At its core, blockchain operates on a decentralized network of computers, or "nodes," eliminating the need for a central authority to oversee transactions. Each node in the network maintains a complete copy of the ledger, known as a distributed ledger, ensuring redundancy and transparency across the system.

Transactions are organized into blocks, with cryptographic hashes linking each block to the one preceding it, creating a chronological and secure record of all network activity. This block-by-block sequence forms the blockchain, providing an immutable and transparent transaction history.

To confirm the current state of the ledger, blockchain employs consensus mechanisms such as proof of stake or proof of work. These mechanisms ensure agreement among network participants regarding the validity of transactions, maintaining the integrity of the ledger.

Furthermore, cryptographic hash functions are utilized to validate the data within each block, ensuring its accuracy and integrity. Any attempt to alter the data within a block would require changing all subsequent blocks, making tampering practically impossible.

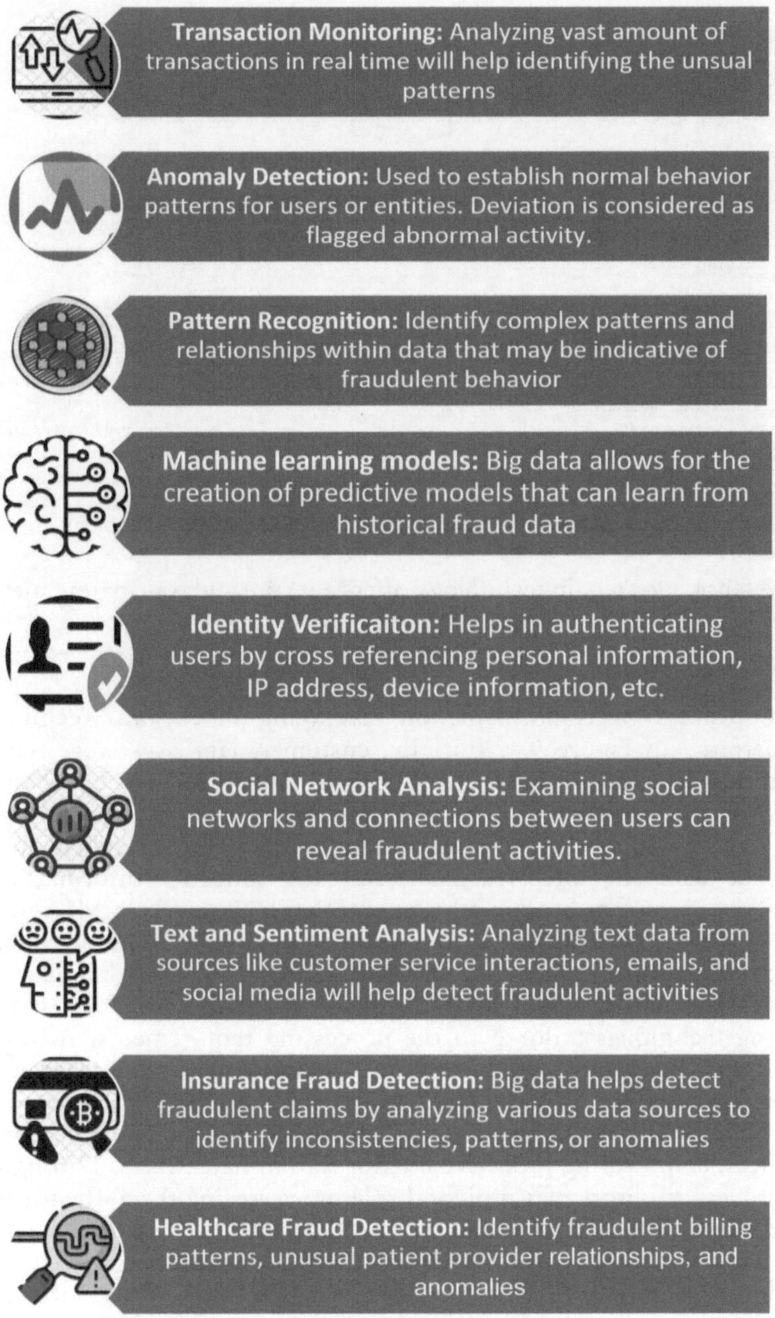

Figure 9.3 Applications of big data in fraud detection.

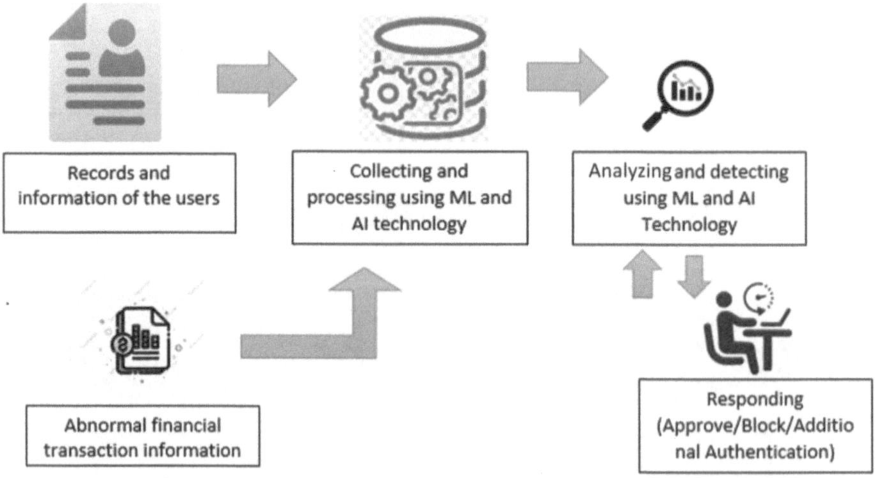

Figure 9.4 Process of verifying the fraud transaction in blockchain technology.

In essence, blockchain technology offers a secure and transparent method for recording and verifying transactions across various industries, promising to revolutionize traditional systems by enhancing trust, efficiency, and accountability.

The process of fraud detection leveraging blockchain technology is illustrated in Figure 9.4. Initially, customer databases store records containing user information, forming the foundation for fraud detection within transactions. These records serve as vital repositories for the data required to identify anomalies and potentially fraudulent activities. The next step involves harnessing the collective information of users for processing through advanced technologies such as ML and AI. This integration allows for a comprehensive analysis of data, enabling the identification of patterns, trends, and irregularities that may signify fraudulent behavior. Subsequently, the data undergoes analysis utilizing techniques tailored to the processing requirements. Advanced algorithms sift through the information, employing sophisticated methodologies to discern fraudulent activities accurately. Through meticulous examination, these techniques enable the prediction of potential fraud instances, empowering proactive measures to mitigate risks. Throughout this process, auditors play a pivotal role in overseeing the authentication procedures. They are responsible for validating transactions, ensuring compliance with established protocols, and providing approval, denial, or additional scrutiny as deemed necessary. In cases where abnormal financial transactions are flagged, auditors are promptly notified to verify the authenticity of the transaction. This iterative approach underscores the importance of continuous monitoring and scrutiny in the realm of fraud detection. By leveraging blockchain technology in conjunction with ML, AI, and rigorous auditing protocols, organizations can bolster their defenses against fraudulent activities, safeguarding assets and maintaining the integrity of financial transactions.

9.4.1 Applications of fraud detection in blockchain technology

Blockchain-based fraud detection has enormous potential to improve security, transparency, and trust in a variety of businesses. The following are some uses of blockchain technology fraud detection.

Blockchain technology offers a wide array of applications for fraud detection across various sectors, revolutionizing the way we approach security and transparency. One of the key areas where blockchain shines is in auditing smart contracts. These self-executing contracts, inherent to blockchain, undergo meticulous scrutiny to uncover security flaws or malicious code. By implementing fraud detection methods, organizations can identify and address potential issues early on, minimizing the risk of exploitation. Moreover, blockchain enhances supply chain transparency by enabling traceability of items from their origin to the end consumer. Fraud detection mechanisms ensure that the data recorded on the blockchain aligns with the actual flow of goods, swiftly identifying any discrepancies or attempts to manipulate information. In the realm of financial services, blockchain simplifies Anti-Money Laundering (AML) and Know Your Customer (KYC) procedures by establishing a decentralized identity verification system. Fraud detection algorithms track and flag suspicious transactions, bolstering security and compliance efforts. Cross-border payments benefit from blockchain's secure and transparent ledger, with fraud detection mechanisms monitoring transactions for anomalies, safeguarding funds and verifying the legitimacy of involved parties. Tokenization and asset management, facilitated by blockchain, rely on fraud detection to monitor ownership and transfer of assets, ensuring authorized transactions and preventing fraud in areas like real estate and art. Decentralized Finance (DeFi) applications utilize fraud detection to scrutinize decentralized exchanges and lending protocols for suspicious activities, unauthorized access, and potential smart contract exploits. Blockchain's secure digital identity verification platform is instrumental in detecting identity theft and fraudulent activities through analysis of identity-related transactions and patterns. In the insurance sector, blockchain records transparent and tamper-proof claims data, while fraud detection algorithms identify irregularities, reducing insurance fraud and streamlining claims processing.

9.4.2 Advantages of BCT in detecting frauds

Blockchain technology offers numerous advantages for detecting and preventing fraud, revolutionizing the way we approach security and accountability in transactions. Some of the key advantages include:

Transparency and Traceability: The inherent nature of blockchain ensures a transparent and auditable trail of transactions. Every transaction is recorded chronologically, providing participants with access to the entire transaction history. This transparency facilitates the detection of fraud by enabling quick identification and investigation of any suspicious or unauthorized activity.

Decentralization: Unlike traditional centralized systems vulnerable to fraud due to a single point of failure, blockchain distributes the ledger across a network of nodes, making it decentralized. This decentralization significantly reduces the risk of fraudulent activity or illegal access, as there is no single point that can be exploited.

Enhanced Security: Blockchain employs advanced cryptographic algorithms to safeguard user identities and transactions through encryption. By bolstering data security and privacy, this cryptographic layer fortifies the blockchain against fraud attempts such as unauthorized access or data breaches.

Real-Time Monitoring and Alerts: Integration of fraud detection algorithms and real-time monitoring tools with blockchain networks enables instant notifications in response to unusual or suspicious activity. This proactive approach empowers organizations to swiftly investigate and take action against potential fraud attempts, minimizing the impact of fraudulent activities. Overall, blockchain technology's transparency, decentralization, security features, and real-time monitoring capabilities make it a powerful tool for detecting and preventing fraud across various industries, ensuring integrity and trust in transactions and processes.

9.4.3 Disadvantages of using blockchain technology in fraud detection

While blockchain technology presents numerous advantages for fraud detection, it also comes with its fair share of challenges and drawbacks that need to be addressed. Some specific disadvantages include:

Limited Regulation and Standards: The absence of standardized regulations and industry standards for blockchain technology creates uncertainties in legal and compliance frameworks. This lack of clear guidelines may impede the adoption of blockchain for fraud detection, particularly in heavily regulated industries where adherence to compliance standards is crucial.

Privacy Concerns: Despite its transparency, blockchain's public nature can raise privacy concerns. Many blockchain networks expose sensitive transaction details to all participants, potentially compromising user privacy. Balancing transparency for fraud detection with the need to protect user privacy poses a significant challenge in blockchain implementation.

Irreversibility of Transactions: Once recorded on the blockchain, transactions are typically irreversible. While immutability is a core strength of blockchain, it becomes a drawback when dealing with erroneous transactions or instances where fraud is detected post-transaction. Unlike traditional systems with chargeback mechanisms, blockchain transactions are final, making rectification challenging.

9.5 FRAUD DETECTION USING ARTIFICIAL INTELLIGENCE

AI has emerged as a crucial tool in the fight against fraud, leveraging advanced algorithms and ML techniques to detect and prevent fraudulent activities. Here are some key factors where AI is utilized in fraud detection:

Anomaly Detection: AI algorithms, particularly unsupervised ML models, establish patterns of normal behavior and flag deviations as anomalies, potentially indicating fraudulent activity. This approach is vital for identifying previously unknown and evolving forms of fraud.

Machine Learning Models: Supervised ML models are trained on historical data, including both legitimate and fraudulent transactions. These models learn to recognize patterns associated with fraud and can predict the likelihood of a transaction being fraudulent in real time, using algorithms like decision trees, random forests, and support vector machines.

Behavioral Analytics: AI-driven behavioral analytics analyze patterns of user behavior across various channels to detect anomalies that may indicate fraudulent activity, especially in online banking and e-commerce. This helps in identifying deviations from established behavioral patterns.

Predictive Analytics: Predictive analytics forecast future events using historical data and statistical algorithms. In fraud detection, predictive models assess the likelihood of a transaction or activity being fraudulent based on patterns identified in historical data, enabling proactive prevention of fraud.

Real-Time Monitoring: AI systems enable real-time monitoring of transactions, user activities, and system logs. By continuously analyzing incoming data, AI algorithms quickly identify and respond to suspicious behavior, reducing the time needed to detect and prevent fraud.

Biometric Authentication: AI-powered biometric authentication methods, such as facial recognition and fingerprint scanning, add an extra layer of security to fraud prevention efforts. These technologies ensure that individuals accessing systems or making transactions are authenticated securely.

Robotic Process Automation (RPA): RPA automates repetitive tasks in fraud detection processes, reducing manual effort and ensuring consistency. By automating routine tasks, RPA speeds up response times and improves fraud detection protocols.

Adaptive Learning: AI systems adapt and evolve based on new data and emerging fraud patterns. This adaptability allows the system to stay effective in dynamic environments where fraudsters continually modify their tactics. The integration of AI in fraud detection not only enhances the accuracy of identifying fraudulent activities but also enables organizations to stay ahead of evolving fraud schemes. As fraudsters become more sophisticated, AI technologies provide valuable tools for creating robust and adaptive fraud detection systems.

9.6 CONCLUSION

In conclusion, fraud detection and financial crime prevention are indispensable components of safeguarding the integrity and security of financial systems. The constantly changing terrain of cyberthreats necessitates the implementation of robust and adaptive measures, including behavioral analysis, anomaly detection, encryption, and incident response plans. Real-world examples, such as the Target data breach, underscore the importance of constant vigilance and the need for financial institutions to stay ahead of sophisticated adversaries. Collaboration with law enforcement, transparent customer communication, and ongoing education efforts contribute to a comprehensive strategy. While no system is completely impervious to attacks, a well-rounded and dynamic approach is essential to mitigate risks, protect customer trust, and maintain the stability of financial ecosystems.

REFERENCES

1. Saxena, S.; Bhushan, B.; Ahad, M. A., Blockchain based solutions to secure IoT: Background, integration trends and a way forward. *Journal of Network and Computer Applications* **2021**, 181, 103050.
2. Aldhaheri, S.; Alghazzawi, D.; Cheng, L.; Barnawi, A.; Alzahrani, B. A., Artificial immune systems approaches to secure the internet of things: A systematic review of the literature and recommendations for future research. *Journal of Network and Computer Applications* **2020**, 157, 102537.

3. Shilpa, B.; Kumar, P. R.; Jha, R. K., LoRa DL: A deep learning model for enhancing the data transmission over LoRa using autoencoder. *The Journal of Supercomputing* **2023**, 79, 17079–17097.
4. Singh, S.; Sharma, P. K.; Yoon, B.; Shojafar, M.; Cho, G. H.; Ra, I. H., Convergence of blockchain and artificial intelligence in IoT network for the sustainable smart city. *Sustainable Cities and Society* **2020**, 63, 102364.
5. Deebak, B. D.; Fadi, A. T., Privacy-preserving in smart contracts using blockchain and artificial intelligence for cyber risk measurements. *Journal of Information Security and Applications* **2021**, 58, 102749.
6. Ranjan, R.; Pandey, D.; Rai, A. K.; Gupta, D.; Singh, P.; Kumar, P. R.; Mohanty, S. N., A manifold-level hybrid deep learning approach for sentiment classification using an autoregressive model. *Applied Sciences* **2023**, 13(5), 3091.
7. Ahanger, T. A., Defense scheme to protect IoT from cyber attacks using AI principles. *International Journal of Computers Communications & Control* **2018**, 13, 915–926.
8. Atlam, H. F.; Walters, R. J.; Wills, G. B., Intelligence of things: Opportunities & challenges. In *Proceedings of the 2018 3rd Cloudification of the Internet of Things (CIoT)*, Paris, France, 2–4 July **2018**, pp. 1–6.
9. Qian, Y.; Jiang, Y.; Chen, J.; Zhang, Y.; Song, J.; Zhou, M.; Pustišek, M., Towards decentralized IoT security enhancement: A blockchain approach. *Computers and Electrical Engineering* **2018**, 72, 266–273.
10. Shen, M.; Tang, X.; Zhu, L.; Du, X.; Guizani, M., Privacy-preserving support vector machine training over blockchain-based encrypted IoT data in smart cities. *IEEE Internet of Things Journal* **2019**, 6, 7702–7712.
11. Arivazhagan, N.; Somasundaram, K.; Mohammad, G. B.; Kumar, P. R.; et al., Cloud-Internet of Health Things (IOHT) task scheduling using hybrid moth flame optimization with deep neural network algorithm for e-healthcare systems. *Scientific Programming* **2022**, 2022, 1–12.
12. Singh, S. K.; Rathore, S.; Park, J. H., Blockiotintelligence: A blockchain-enabled intelligent IoT architecture with artificial intelligence. *Future Generation Computer Systems* **2020**, 110, 721–743.
13. Han, X.; Zhang, R.; Liu, X.; Jiang, F., Biologically inspired smart contract: A blockchain-based DDoS detection system. In *Proceedings of the 2020 IEEE International Conference on Networking, Sensing and Control (ICNSC)*, Nanjing, China, 30 October–2 November **2020**, pp. 1–6.
14. Mohammad, G. B.; Selvarajan Shitharth; Kumar, P. R., Integrated machine learning model for an URL phishing detection. *International Journal of Grid and Distributed Computing* **2021**, 14(1), 513–529.
15. Sandner, P.; Gross, J.; Richter, R., Convergence of blockchain, IoT, and AI. *Frontiers in Blockchain* **2020**, 3, 522600.
16. Rodrigues, B.; Bocek, T.; Lareida, A.; Hausheer, D.; Rafati, S.; Stiller, B., A blockchain-based architecture for collaborative DDoS mitigation with smart contracts. In *Proceedings of the IFIP International Conference on Autonomous Infrastructure, Management and Security*, Zurich, Switzerland, 10–13 July **2017**, Springer: Cham, Switzerland, pp. 16–29.
17. Rathore, S.; Kwon, B. W.; Park, J. H., BlockSecIoTNet: Blockchain-based decentralized security architecture for IoT network. *Journal of Network and Computer Applications* **2019**, 143, 167–177.

18. Wu, B.; Li, Q.; Xu, K.; Li, R.; Liu, Z., Smartretro: Blockchain-based incentives for distributed IoT retrospective detection. In *Proceedings of the 2018 IEEE 15th International Conference on Mobile Ad Hoc and Sensor Systems (MASS)*, Chengdu, China, 9–12 October **2018**, pp. 308–316.
19. Kumar, P. R.; Ananthan, T., Machine vision using LabVIEW for label inspection. *Journal of Innovation in Computer Science and Engineering (JICSE)* **2019**, 9(1), 58–62.
20. Talukder, S.; Roy, S.; Al Mahmud, T., An approach for an distributed anti-malware system based on blockchain technology. In *Proceedings of the 2019 11th International Conference on Communication Systems & Networks (COMSNETS)*, Bengaluru, India, 7–11 January **2019**, pp. 1–6.
21. Kumar, P. R., Wireless mobile charger using inductive coupling. *Journal of Emerging Technologies and Innovative Research (JETIR)* **2018**, 5(10), 40–44.
22. Li, W.; Tug, S.; Meng, W.; Wang, Y., Designing collaborative blockchained signature-based intrusion detection in IoT environments. *Future Generation Computer Systems* **2019**, 96, 481–489.
23. Spathoulas, G.; Giachoudis, N.; Damiris, G. P.; Theodoridis, G., Collaborative blockchain-based detection of distributed denial of service attacks based on Internet of Things botnets. *Future Internet* **2019**, 11, 226.
24. Shilpa, B.; Kumar, P. R.; Jha, R. K., Spreading factor optimization for interference mitigation in dense indoor LoRa networks. *IEEE IAS Global Conference on Emerging Technologies (GlobConET)*, London, UK, **2023**, pp. 1–5.
25. Cheema, M. A.; Qureshi, H. K.; Chrysostomou, C.; Lestas, M., Utilizing blockchain for distributed machine learning based intrusion detection in internet of things. In *Proceedings of the 2020 16th International Conference on Distributed Computing in Sensor Systems (DCOSS)*, Marina del Rey, CA, 25–27 May **2020**, pp. 429–435.
26. Aldhaheri, S.; Alghazzawi, D.; Cheng, L.; Alzahrani, B.; Al-Barakati, A., Deep DCA: Novel network-based detection of IoT attacks using artificial immune system. *Applied Sciences* **2020**, 10, 1909.
27. Nespoli, P.; Mármol, F. G.; Vidal, J. M., A bio-inspired reaction against cyberattacks: AIS-powered optimal countermeasures selection. *IEEE Access* **2021**, 9, 60971–60996.
28. Kumar, P. R., Position control of a stepper motor using LabVIEW. *3rd International Conference on Recent Trends in Electronics, Information Communication Technology (RTEICT)*, Bangalore, India, May **2018**, pp. 1551–1554.
29. Ghali, A. A.; Ahmad, R.; Alhussian, H., A framework for mitigating DDoS and DOS attacks in IoT environment using hybrid approach. *Electronics* **2021**, 10, 1282.
30. Kumar, P.; Kumar, R.; Gupta, G. P.; Tripathi, R., A distributed framework for detecting DDoS attacks in smart contract-based Blockchain-IoT systems by leveraging Fog computing. *Transactions on Emerging Telecommunications Technologies* **2021**, 32, e4112.

Chapter 10

Machine learning techniques for blockchain technology

A review of recent advances and unresolved issues

I. Sheik Arafat, S. Karthiyayini, S. M. Haji Nishath, and R. Karthikeyan

10.1 INTRODUCTION TO MACHINE LEARNING AND DEEP LEARNING

10.1.1 Fundamentals of ML/DL

The concepts, techniques, and methodologies that are essential for understanding and harnessing the power of artificial intelligence encompass a vast and complex landscape within deep learning (DL) and machine learning (ML) [9]. The fundamental principle behind ML is to allow computer systems to learn from data without the need for explicit programming. The two primary forms of learning are unsupervised learning, which is the process of drawing patterns and structures from unlabeled data, and supervised learning, in which algorithms learn from labeled instances to generate predictions or judgments. Another paradigm, reinforcement learning, emphasizes learning through interaction with an environment, with rewards guiding the learning process. Because DL, a subset of ML, can automatically extract hierarchical representations from data, it has emerged as a major player in the field. The core elements of DL are neural networks, which replicate the networked organization of neurons in the human brain. They are made up of node layers that process and alter data. Activation functions introduce nonlinearity into neural network architectures, enhancing their ability to model complex relationships in data. Training neural networks involves iteratively optimizing through backpropagation, where gradients are computed and used to update model parameters, minimizing the difference between predicted and actual outputs.

Recurrent neural networks (RNNs) are adept at processing sequential data, which makes them appropriate for tasks like natural language processing (NLP) and time series prediction. DL architectures, such as convolutional neural networks (CNNs), thrive in tasks like image recognition and computer vision by utilizing spatial hierarchies. Regularization strategies like dropout reduce overfitting and guarantee generalization to new data, while optimization approaches like gradient descent and its variants adjust model parameters. Data preprocessing is a crucial step in preparing data for

DOI: 10.1201/9781032663005-10

ML/DL models. It entails cleaning, manipulating, and encoding data. Evaluation and validation techniques are used to assess model performance, guiding the selection of appropriate algorithms and hyperparameters. Ethical considerations are significant in the ML/DL landscape, with concerns about bias, fairness, and transparency leading to calls for responsible AI development. As the field progresses, advanced topics like generative adversarial networks (GANs) and reinforcement learning offer new possibilities, while ongoing research and collaboration contribute to innovative applications and societal impact, shaping a future where ML/DL continue to shape our world.

10.1.2 Overview of neural networks and deep learning architectures

DL is based on neural networks, which offer a computational model inspired by the composition and operations of the human brain. Artificial neurons, which are interconnected nodes arranged into layers to process and manipulate incoming data, make up these networks. This approach is expanded upon by numerous layers in DL architectures, which enable the automatic extraction of hierarchical representations from data. Information travels from input nodes to output nodes in a conventional feedforward neural network, passing via hidden layers that perform a series of weighted modifications to the input. The network may learn intricate correlations in the data by introducing nonlinearity through activation functions. Convolutional layers are used by CNNs to extract spatial hierarchies of features, which allows CNNs to handle grid-like data, such photographs. Tasks like semantic segmentation, object detection, and picture classification have been revolutionized by these systems. RNNs, however, are made especially for processing sequential input. RNNs use directed cycles formed by node connections to capture temporal dependencies. RNNs are highly effective at tasks like speech recognition, NLP, and time series prediction because of their capacity to handle inputs of different durations and model sequences. An alternative to RNNs, long short-term memory (LSTM) networks solve the vanishing gradient problem by implementing gated mechanisms that control the information flow over time. These architectures are now essential for activities like emotion analysis and language translation that call for the preservation of memory and context. In conclusion, advances in a variety of fields have been fueled by the flexibility and adaptability of neural networks and DL architectures, which have sparked creativity and revolutionized the artificial intelligence sector.

10.1.3 Applications of ML/DL in various industries

ML and DL have been widely adopted in various industries, providing innovative solutions to intricate problems and driving efficiency and optimization across a diverse range of sectors.

In the field of healthcare, ML/DL applications encompass a wide range of uses, from the analysis of medical images for disease diagnosis to predictive analytics for patient outcomes and personalized treatment plans. ML models examine extensive medical data to detect patterns and forecast the progression of diseases, leading to a transformative impact on healthcare delivery and improved patient outcomes.

Financial institutions use ML and DL for algorithmic trading, risk management, and fraud detection. Massive datasets are combed through by ML algorithms to spot fraudulent activity in financial transactions, and risk management models assess market and credit risks to help in decision-making. Algorithmic trading strategies employ DL techniques to analyze market data and execute trades automatically, thereby optimizing investment portfolios and maximizing returns.

Retailers employ ML/DL for personalized recommendations, inventory management, and dynamic pricing. Recommendation systems analyze customer behavior to provide tailored product suggestions, while ML models forecast demand and optimize inventory levels to prevent stockouts and reduce costs. In order to increase sales and profitability, dynamic pricing algorithms modify prices in real time in response to consumer preferences and market conditions.

ML/DL is used by manufacturing organizations for supply chain optimization, quality control, and predictive maintenance. While DL algorithms check items for flaws and deviations from quality standards, predictive maintenance models use sensor data from equipment to anticipate breakdowns and reduce downtime. ML techniques optimize supply chain operations by forecasting demand, managing inventory, and improving logistics efficiency.

In the transportation sector, ML/DL powers autonomous vehicles, optimizes route planning, and enhances public transit management. Self-driving cars rely on DL algorithms to process sensor data and make real-time driving decisions, while ML models optimize transportation routes based on traffic patterns and weather conditions. Public transit systems benefit from ML techniques for predicting passenger demand, optimizing schedules, and improving service reliability.

Across the energy sector, ML/DL applications encompass predictive maintenance for infrastructure, energy consumption forecasting, and the optimization of smart grids. ML models predict equipment failures in energy infrastructure, optimize energy generation and distribution, and balance supply and demand on smart grids, contributing to sustainability and efficiency in energy management.

As seen in Figure 10.1, the assistance of DL in image analysis and recognition makes it possible to identify objects, people, or any actions in a picture.

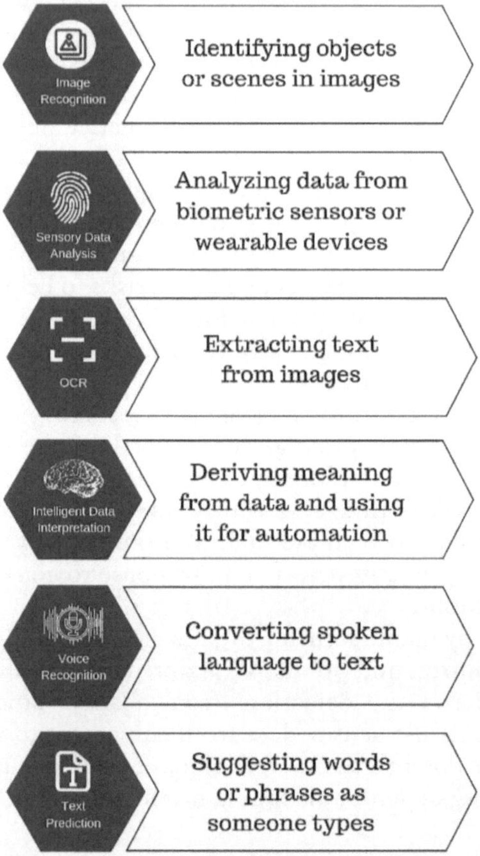

Figure 10.1 Key benefits of deep learning techniques in various fields [1].

10.2 FOUNDATIONS OF BLOCKCHAIN TECHNOLOGY

10.2.1 Basics of blockchain technology

The foundational ideas of blockchain technology lay the framework for understanding its decentralized and immutable characteristics, which are essential for its widespread adoption and acceptance. Fundamentally, a blockchain operates as a distributed ledger that documents transactions throughout a network of linked nodes, guaranteeing trust, security, and transparency without the need for middlemen. Every block in the blockchain consists of a collection of transactions that are cryptographically connected to the block before it, creating a sequential arrangement of blocks. Through the use of cryptographic hashing algorithms, this interlinking creates an immutable record of transactions, making it extremely impossible to alter previous data without the approval of the majority of network users [4].

Blockchain operates on a network architecture that is decentralized, whereby every participant, or node, maintains a complete copy of the entire blockchain ledger. This decentralization guarantees resilience against singular points of failure and prevents unauthorized manipulation of data, thereby enhancing the security and dependability of the network.

By facilitating agreement among network participants regarding the authenticity of transactions, consensus mechanisms are essential to maintaining the integrity of the blockchain. Two popular consensus techniques are Proof of Stake (PoS) and Proof of Work (PoW). In order to validate transactions and add new blocks to the blockchain, PoW requires users to solve complex mathematical puzzles, whereas PoS chooses block validators based on stake and rewards honest behavior with money.

Another essential component of blockchain technology are smart contracts, which are self-executing agreements with predefined terms and conditions built into the network. These programmable contracts eliminate the need for middlemen and streamline a variety of corporate operations by automatically carrying out activities when predetermined circumstances are satisfied.

Blockchain technology has numerous uses in a variety of sectors, such as supply chain management, healthcare, and finance. It makes transactions safer and more transparent, lowers the risk of fraud and error, and promotes the development of decentralized apps (DApps) and new business models. To fully realize the disruptive potential of blockchain technology and explore its many applications in the modern digital economy, one must acquire a thorough understanding of its fundamentals. Figure 10.2 highlights the distinctive advantages of blockchain in comparison to other technologies.

10.2.2 Types of blockchains

Blockchain technology has developed to accommodate various use cases, resulting in different categories of blockchains. These categories can be broadly classified into three main groups: public blockchains, private blockchains, and consortium blockchains. Each group possesses its own distinct characteristics, advantages, and use cases.

10.2.2.1 Public blockchains

Public blockchains are decentralized networks in which anybody may join, see, edit, and verify transactions without needing to obtain authorization. Anyone can access the ledger and participate in the consensus process thanks to these open and transparent blockchains. Blockchains that are publicly accessible include Ethereum and Bitcoin.

Public blockchains operate on the foundations of immutability, transparency, and decentralization. Since everyone in the network has equal access

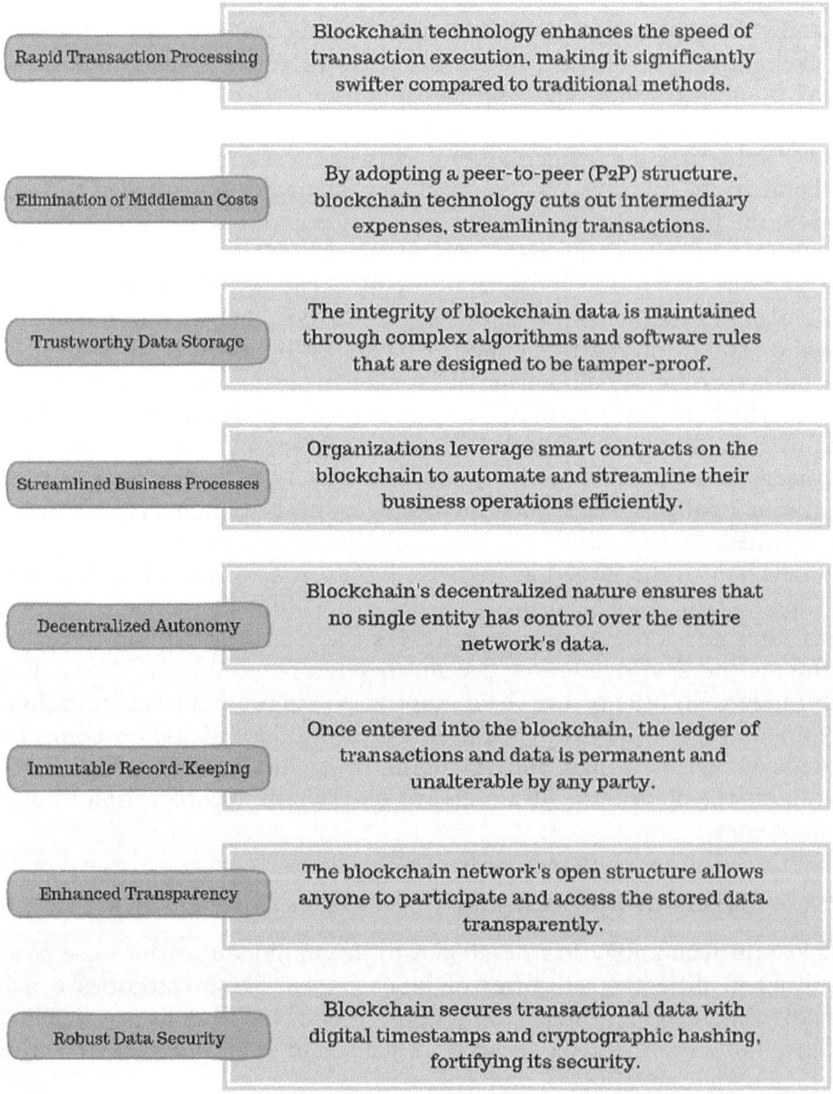

Figure 10.2 Benefits of blockchain compared to other technologies [1].

to the ledger and rights, there can be no censorship or unreliable transactions. To confirm transactions and safeguard the network, public blockchains rely on consensus techniques like PoW or PoS.

One of the primary advantages of public blockchains is their high level of security, achieved through decentralized consensus and cryptographic hashing. Once confirmed, transactions on public blockchains are irreversible, offering robust guarantees of immutability and resistance to tampering.

Public blockchains are particularly well-suited for applications that necessitate a high level of security, censorship resistance, and transparency. They are commonly utilized for cryptocurrency transactions, decentralized finance (DeFi), token issuance, and decentralized applications (DApps).

10.2.2.2 Private blockchains

A centralized network where participation and access are limited to authorized entities is known as a private blockchain, also known as a permissioned blockchain. Private blockchains require authorization to join, read, and alter data, in contrast to public blockchains that are accessible to everybody. Examples of well-known private blockchain platforms are R3 Corda and Hyperledger Fabric.

Private blockchains are usually used in consortiums or organizations with well-known and reliable members. Only authorized organizations are able to access sensitive data on the network thanks to permission systems and access controls.

While private blockchains sacrifice decentralization and censorship resistance in favor of improved performance and scalability, they offer benefits such as enhanced privacy, control, and efficiency. Participants can engage in trusted transactions with one another without exposing sensitive information to the public.

Enterprise applications where privacy, compliance, and scalability are critical, like supply chain management, identity verification, and document management, are ideal candidates for private blockchain technology.

10.2.2.3 Consortium blockchains

A hybrid approach that combines the best features of public and private blockchains is represented by consortium blockchains. A pre-selected set of participants manages the network collaboratively, sharing authority and decision-making duties in a consortium blockchain. Consortium blockchains enforce permissioning rules and access limitations while keeping a certain degree of decentralization.

Consortium blockchains are overseen by a number of reliable organizations, in contrast to private and public blockchains, which are accessible to everyone. These groups work together to uphold consensus rules, validate transactions, and keep the blockchain ledger current.

Consortium blockchains are appropriate for use cases requiring several stakeholders to work together and safely share data because they provide a balance between decentralization and control. Interbank payment systems, industry consortia, and supply chain networks are a few examples. Benefits of consortium blockchains include enhanced productivity, lower costs, and greater participant trust. Consortiums can improve transparency, expedite business procedures, and provide new value offerings for their members by utilizing blockchain technology.

10.3 TAXONOMY OF BLOCKCHAIN-BASED DEEP LEARNING FRAMEWORKS

This taxonomy discusses the organization of blockchain-based DL frameworks. DL algorithms and blockchain technology combine to provide a number of benefits, such as automated decision-making, data protection, precise forecasting, effective data market management, and enhanced system resilience. In this part, a thematic taxonomy is presented to classify, according to various criteria, the unionization of DL techniques and blockchain. These elements, as seen in Figure 10.3, draw attention to the parallels and discrepancies between the cutting-edge DL frameworks based on blockchain technology. An overview of the chosen parameters and their technical details is given in the section that follows.

10.4 SMART CONTRACTS AND DECENTRALIZED APPLICATIONS (DAPPS)

In the realm of blockchain technology, smart contracts and decentralized applications (DApps) are essential components that revolutionize the way transactions are carried out and applications are developed on

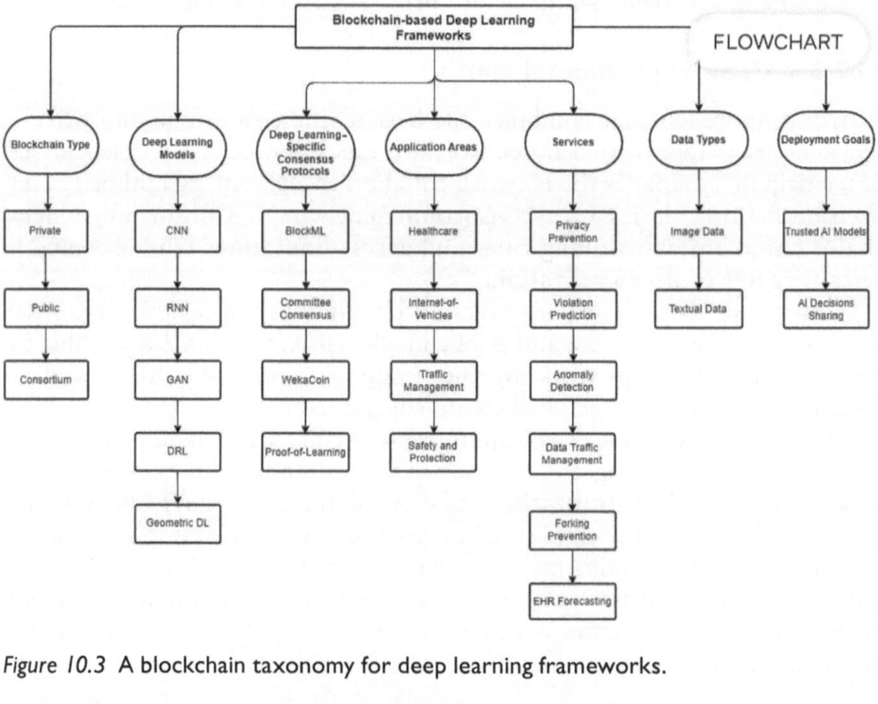

Figure 10.3 A blockchain taxonomy for deep learning frameworks.

decentralized networks. Smart contracts are self-executing agreements that are encoded in the blockchain with predetermined terms and circumstances. They ensure transparent and trustless transactions by automating and enforcing the terms of an agreement between parties, doing away with the need for middlemen.

Smart contracts are written in programming languages like Chaincode for Hyperledger Fabric and Solidity for Ethereum, and they are then implemented on blockchain platforms. Smart contracts offer a dependable method of carrying out agreements without the possibility of fraud or manipulation once they are deployed. Once deployed, they become immutable and cannot be changed. Smart contracts are triggered by predefined conditions and execute automatically, facilitating the exchange of assets, information, or services between parties.

Applications that run on decentralized networks and lack central authority or control are known as decentralized apps (DApps). These apps make use of blockchain technology and smart contracts. DApps are made to be transparent, open-source, and censorship-resistant so that everyone can use them and communicate with them without authorization.

DApps can be created for a wide range of use cases in a variety of sectors, including social media, gaming, banking, and supply chain management. They are intrinsically safe and reliable because they inherit the decentralized, immutable, and transparent characteristics of blockchain technology.

The capacity of smart contracts and DApps to automate complicated procedures and remove middlemen, which lowers costs and boosts efficiency, is one of its main benefits. Smart contracts have the potential to automate the execution of financial agreements, such as insurance policies, derivatives, and loans, in the finance industry. This can be achieved without the need for traditional intermediaries like banks or insurance firms. This expedites the procedure and lowers the possibility of mistakes and disagreements.

Additionally, through tokenization and DeFi, smart contracts and DApps make it possible to develop new business models and revenue streams. On the blockchain, tokens stand for digital assets or rights, and they can be used to speed up transactions, grant access to services, or take part in DApp governance processes. Without depending on conventional financial institutions, DeFi protocols use smart contracts to provide decentralized financial products and services like lending, borrowing, trading, and yield farming.

DApps and smart contracts do have some difficulties, though. Because security flaws in smart contract programming might result in money losses and exploits, it is crucial to have strict testing and auditing procedures. Scalability and usability issues also pose challenges to the widespread adoption of DApps, as blockchain networks struggle to handle large transaction volumes and provide seamless user experiences.

10.5 CHALLENGES AND OPPORTUNITIES IN BLOCKCHAIN INTELLIGENCE

10.5.1 Analyzing the complexities of blockchain data

Jan et al. discuss the importance of analyzing blockchain data and identifies four aspects of analysis: security, privacy, performance, and price prediction. However, it does not specifically mention analyzing the complexities of blockchain data.

Analyzing the complexities of blockchain data involves navigating through the vast amount of information stored on the blockchain, understanding its structure, and extracting valuable insights to derive meaningful conclusions. Blockchain data is inherently complex due to its decentralized and distributed nature, as well as its cryptographic mechanisms that ensure security and immutability.

One of the primary challenges in analyzing blockchain data is its sheer volume and variety. Blockchains continuously grow as new transactions are added to the ledger, resulting in massive datasets that require efficient storage, retrieval, and processing mechanisms. Moreover, blockchain data comes in various formats, including transaction records, smart contract code, and metadata, which adds to the complexity of analysis.

Another complexity arises from the transparency and pseudonymity of blockchain transactions. While blockchain data is transparent and publicly accessible, identifying the entities behind transactions can be challenging due to the use of cryptographic addresses. Analyzing transaction patterns and clustering techniques can help attribute transactions to specific entities or addresses, providing insights into user behavior and network activity.

Understanding the underlying consensus processes and protocol guidelines that control the blockchain network is also necessary for blockchain data analysis. Different consensus techniques, such as PoW, PoS, or delegated Proof of Stake (dPoS), may be used by different blockchains. Each has its own set of guidelines and motivations. Understanding these consensus techniques is necessary for analyzing blockchain data in order to decipher transaction confirmations, block propagation delays, and network security.

Furthermore, blockchain data analysis often involves exploring the temporal aspects of transactions, such as timestamps and block intervals. Analyzing transaction timestamps can reveal patterns of activity, transaction frequency, and time-sensitive events on the blockchain. Understanding block intervals and confirmation times is essential for assessing network performance, scalability, and reliability.

Researchers and analysts use a range of tools and methods, such as data visualization, statistical analysis, ML, and network analysis, to analyze blockchain data efficiently. Blockchain data may be visualized and patterns and trends can be found with the aid of data visualization techniques like graphs, charts, and heatmaps. Quantitative insights into blockchain metrics and performance indicators can be obtained through statistical analysis approaches including time series analysis and descriptive statistics.

ML algorithms can be used to identify abnormalities or fraudulent activity, forecast future trends, and find hidden patterns and correlations within blockchain data. Blockchain network architecture and structure are analyzed, with the use of network analysis techniques like graph theory and centrality measurements that help discover important individuals, communities, and network dynamics.

10.5.2 Identifying security and privacy challenges

Identifying the challenges pertaining to security and privacy in blockchain technology is of utmost importance in order to ensure the integrity, confidentiality, and resilience of systems that are based on blockchain. Although blockchain offers numerous advantages, such as decentralization, transparency, and immutability, it also presents distinctive considerations with regards to security and privacy that must be effectively addressed in order to mitigate risks and vulnerabilities.

The possibility of 51% assaults, in which a single entity or a group of entities seizes control of the majority of the mining power within the network, is one of the main security challenges facing blockchain technology. When a 51% attack occurs on a blockchain that uses the PoW consensus mechanism, like Bitcoin, the attacker may be able to modify transaction confirmations, spend coins twice, or even censor transactions. Similar to this, a majority stakeholder in PoS blockchains has the ability to influence the consensus process, hence jeopardizing network security.

In essence, smart contracts are self-executing contracts with specified rules stored into the blockchain. As such, they are susceptible to malicious attacks, logical fallacies, and coding faults. By taking advantage of smart contract flaws, one can steal digital assets, incur financial losses, or even cause disruptions to blockchain-based decentralized applications, or DApps.

Another major issue with blockchain technology is privacy, especially in public blockchains where transaction data is clear and available to all network users. Despite the fact that blockchain transactions are pseudonymous—that is, they are associated with cryptographic addresses rather than actual identities—it is nevertheless possible to track and examine transaction patterns to unearth private data about users' financial activity.

Moreover, data recorded on the blockchain cannot be removed or changed once it has been recorded there, making it unchangeable. Although this immutability guarantees the blockchain data's integrity and resistance to manipulation, it also poses problems for privacy protection. This is due to the possibility that private data saved on the blockchain could remain unchanged for an extended period of time, putting users at risk for privacy violations and problems with regulatory compliance.

Various solutions and best practices are being actively explored by researchers, developers, and politicians to address these security and privacy challenges. Enhancing consensus processes and blockchain protocols is one way to increase scalability and security while lowering the possibility of 51% attacks and other network problems. For example, the use of sophisticated cryptographic methods, like homomorphic encryption or zero-knowledge proofs (ZKPs), can improve privacy by allowing private transactions to occur on public blockchains.

Additionally, developers are focusing on enhancing the security of smart contracts through formal verification, code auditing, and the establishment of bug bounties in order to identify and tackle vulnerabilities before the deployment of these contracts. Through the conduct of comprehensive security assessments and the implementation of robust coding practices, developers can reduce the risk of smart contract exploits and safeguard the digital assets of users against theft or manipulation.

Furthermore, the development of privacy-enhancing technologies, such as privacy coins, mixers, and decentralized identity solutions, is underway with the aim of bolstering privacy protection on public blockchains. The aims of these technologies are to provide users more control on their private information, mask user identities, and anonymize transaction data.

10.5.3 Exploring opportunities for ML/DL integration

Exploration of the potential for the integration of ML and DL with various industries and applications holds substantial promise for driving innovation, enhancing efficiency, and uncovering novel insights. The incorporation of ML/DL techniques with existing systems and processes presents numerous opportunities for optimization, automation, and intelligence across diverse domains.

Li et al. [3] examine the potential of combining data integration methods with ML to enhance model accuracy and automate data transformation workflows.

A prominent prospect is to apply ML and DL to evaluate and extract knowledge from enormous amounts of data produced by companies, institutions, and Internet of Things devices [5]. Massive datasets may be efficiently examined by ML algorithms to find patterns, trends, and correlations that human analysts would not see right away. Organizations may extract meaningful insights and reveal complex relationships from unstructured data sources like text and videos by utilizing DL techniques like neural networks.

Furthermore, by offering predictive analytics and prescriptive recommendations based on historical data and real-time inputs, the integration of

ML/DL can improve decision-making processes. Utilizing ML algorithms, predictive models are able to predict future trends, predict customer behavior, and optimize resource allocation. This gives organizations the ability to make data-driven decisions and acquire a competitive edge in ever-changing markets.

Furthermore, the integration of ML/DL offers opportunities for automation and process optimization across various industries. Robotic process automation (RPA) fueled by ML algorithms can streamline repetitive tasks, reduce manual effort, and enhance operational efficiency in domains such as finance, customer service, and supply chain management. DL techniques such as computer vision and NLP enable the automation of tasks that necessitate comprehension and processing of visual or textual information, such as image recognition, document processing, and language translation.

Moreover, the integration of ML/DL can bolster cybersecurity measures by promptly detecting and mitigating threats. ML algorithms possess the capability to analyze network traffic, identify abnormal behavior, and detect security breaches or intrusions, thereby enabling organizations to proactively defend against cyberattacks and safeguard sensitive data and assets.

In the healthcare sector, the integration of ML/DL holds the potential to revolutionize patient care, disease diagnosis, and drug discovery. ML algorithms can analyze medical imaging data to aid radiologists in the detection of abnormalities and early-stage disease diagnosis. DL techniques such as DL-based drug discovery platforms can expedite the identification of innovative drug candidates and optimize drug development processes, leading to swifter and more efficacious treatments for various medical conditions.

Moreover, ML/DL integration with IoT devices and sensor networks makes it easier to create autonomous, intelligent systems that can react to changing environments. In order to forecast maintenance needs, optimize energy use in smart buildings and industrial facilities, and monitor equipment performance, ML algorithms can evaluate sensor data in real time. IoT sensor data can be processed by DL techniques like CNNs and RNNs to identify anomalies, forecast failures, and improve overall system dependability.

10.6 DATA PREPROCESSING AND FEATURE ENGINEERING FOR BLOCKCHAIN ANALYSIS

10.6.1 Data collection and preprocessing techniques

Data collection and preprocessing methods play a pivotal role in guaranteeing the quality, dependability, and usability of data for ML and data analysis tasks. Proficient data collection necessitates the accumulation of pertinent data sources in a structured and methodical manner, while preprocessing methods concentrate on cleansing, transforming, and preparing the data for analysis. An in-depth examination of data collection and preprocessing methods is presented below:

Data Collection: Identification of Data Sources: The primary step in data collection is the identification and selection of relevant data sources that contain the necessary information for analysis. This may involve databases, APIs, web scraping, IoT devices, sensors, social media platforms, or exclusive data sources.

Data Acquisition: Once the data sources have been identified, data is obtained through various means such as file downloads, database queries, API access, or real-time data collection from sensors or IoT devices. It is crucial to ensure that the collected data is accurate, complete, and representative of the problem domain.

Data Storage: After data collection, it is stored in an appropriate format and data storage system, such as relational databases, NoSQL databases, data lakes, or cloud storage platforms. Considerations for data storage include scalability, reliability, security, and compliance with data privacy regulations.

Data Preprocessing: Data cleaning is the process of locating and fixing mistakes, inconsistencies, and missing values in a dataset. Duplicate record removal, imputation or deletion of missing data handling, data formatting error correction, and treatment of outliers or anomalies are some examples of what this might involve.

Data Transformation: To turn unprocessed data into a format that can be analyzed, data transformation techniques are used. To make sure that features have comparable sizes and distributions, this may involve feature scaling, normalization, or standardization. Categorical variable encoding, data distribution transformations, and feature engineering are examples of additional transformations.

Feature Selection: To determine the most important characteristics or variables that enhance the model's prediction ability, feature selection techniques are applied. This helps prevent overfitting, improve model performance, and decrease dimensionality. Filter, wrapper, and integrated feature selection techniques are based on statistical testing, model performance, or domain expertise.

Data Integration: To establish a single dataset for analysis, data from many sources or datasets are combined. This could mean connecting datasets using relational or geographic operations, aggregating data at different granularities, or merging databases based on shared identities or keys.

Data Sampling: Data sampling techniques are employed to select a subset of data points from the original dataset for analysis. This is especially advantageous for large datasets where processing the entire dataset is computationally expensive or impractical. Common sampling methods include random sampling, stratified sampling, and oversampling/undersampling for imbalanced datasets. By employing effective data collection and preprocessing techniques, data scientists and analysts can ensure the accuracy, reliability, and suitability of the data used for ML and data analysis tasks, thus enabling the generation of meaningful insights and the construction of robust models. These techniques establish the groundwork for successful data-driven decision-making and enable organizations to extract actionable insights from their data assets.

10.6.2 Feature engineering approaches for blockchain data

Feature engineering plays a crucial role in extracting meaningful insights and improving the performance of ML models when dealing with blockchain data. Blockchain data, with its unique characteristics such as transactional nature, timestamped records, and cryptographic security, requires specialized feature engineering approaches to effectively capture relevant information and patterns. Here are some key feature engineering approaches for blockchain data:

Transaction-Level Features: At the most granular level, features can be extracted directly from individual transactions recorded on the blockchain. These features may include transaction amount, transaction type (e.g., payment, transfer, contract execution), sender and receiver addresses, transaction fees, and timestamps. Additionally, derived features such as transaction volume, frequency, and velocity can provide insights into transactional behavior and network activity.

Address-Level Features: Features can also be derived from addresses involved in transactions, including sender and receiver addresses as well as smart contract addresses. Address-level features may include the number of transactions associated with an address, the total value transacted, the number of unique counterparties, and the degree of activity over time. Address clustering techniques can be applied to group addresses belonging to the same entity or entity cluster, enabling more accurate modeling of user behavior.

Temporal Features: Given the timestamped nature of blockchain transactions, temporal features can capture patterns and trends over time. These features may include transaction frequency and volume trends, time intervals between transactions, time-of-day patterns, and cyclical patterns such as weekly or monthly trends. Temporal features enable models to capture seasonality, periodicity, and other time-dependent dynamics in blockchain data.

Jeyakumar et al. propose an automated feature engineering technique for extracting numerous properties from blockchain transactions in order to detect suspicious activity. Graph-based Features: Blockchain data can be represented as a graph, with nodes representing addresses and edges representing transactions between addresses. Graph-based features leverage the structural properties of the transaction graph to capture network topology, connectivity patterns, and centrality measures. Features such as node degree, centrality measures (e.g., betweenness centrality, PageRank), and graph motifs can provide insights into the network structure and identify influential entities or addresses.

Transaction Sequence Features: Sequential patterns in transaction data can be captured using sequence-based features. These features may include transaction sequences or sequences of addresses involved in transactions, which can be encoded using techniques such as sequence embedding or RNNs. Sequence-based features enable models to capture temporal dependencies and behavioral patterns in transaction sequences.

Transaction Metadata Features: Many blockchain platforms support additional metadata fields associated with transactions, such as transaction tags, memos, or smart contract parameters. These metadata fields can provide contextual information about transactions, such as transaction purposes, transaction descriptions, or smart contract parameters. Extracting and encoding metadata features can enrich the feature space and provide additional context for modeling transaction behavior.

10.6.3 Handling imbalanced data and missing values

Handling imbalanced data and missing values is of utmost importance in order to ensure the effectiveness and reliability of ML models. Imbalanced data arises when one class or category in the dataset is significantly more prevalent than others, resulting in biased predictions made by the model. In a similar vein, the presence of missing values in the dataset can introduce both noise and bias into the model, thereby impacting its performance and accuracy.

To tackle imbalanced data, there exist various techniques that can be utilized. One commonly employed approach is resampling, which entails either oversampling the minority class or undersampling the majority class with the aim of achieving a balanced dataset. Oversampling techniques involve duplicating samples from the minority class or generating synthetic samples using methods like SMOTE (Synthetic Minority Oversampling Technique). However, undersampling methods randomly eliminate samples from the majority class in order to match the size of the minority class. Another approach is to adjust the class weights during model training so as to penalize misclassifications of the minority class more heavily, thereby ensuring that the model learns to prioritize correct predictions for both classes equally.

Addressing missing values necessitates meticulous preprocessing. One strategy is to impute the missing values using statistical techniques such as mean, median, or mode imputation, wherein the missing values are replaced with the mean, median, or mode of the respective feature. Alternatively, missing values can be imputed by employing predictive models that are trained on the remaining data, wherein the missing values are predicted based on the observed values of other features. Another approach is to treat missing values as a distinct category or to leverage domain knowledge to infer the missing values based on contextual information.

10.7 FRAUD DETECTION AND ANOMALY DETECTION

10.7.1 Detecting fraudulent activities in blockchain transactions

Ogundokun et al. discuss the use of DL methods to detect phishing attacks in a blockchain transaction network, but it does not specifically mention detecting fraudulent activities in blockchain transactions.

Detecting fraudulent activities in blockchain transactions is a significant challenge due to the decentralized and pseudonymous nature of blockchain networks. While blockchain technology provides transparency and immutability, malicious actors may exploit vulnerabilities or engage in illicit activities such as money laundering, fraud, or theft. To address these threats, various techniques and approaches are utilized to effectively detect and prevent fraudulent transactions.

Anomaly detection is a common method for identifying fraudulent activities in blockchain transactions. Anomaly detection algorithms are capable of identifying transactions or patterns that deviate significantly from the expected behavior of legitimate transactions. These anomalies may manifest as sudden spikes or drops in transaction volume, unusual transaction amounts, irregular transaction patterns, or unexpected changes in network activity. ML algorithms, including clustering, classification, or outlier detection models, can be trained on historical transaction data to identify suspicious patterns and highlight potentially fraudulent transactions for further examination.

Another approach is behavior-based analysis, which involves monitoring and analyzing the behavior of entities (e.g., addresses, users, or smart contracts) on the blockchain network. By tracking transaction history, frequency, volume, and interactions between entities, behavior-based analysis can identify abnormal or suspicious behavior indicative of fraudulent activities. Graph-based techniques, such as network analysis and centrality measures, can be used to identify suspicious entities, detect transaction laundering, and uncover hidden connections between fraudulent actors.

Additionally, pattern recognition techniques can be employed to identify known fraud patterns or signatures in blockchain transactions. These patterns may include common tactics used in fraudulent schemes, such as Ponzi schemes, pump-and-dump schemes, or phishing attacks. By analyzing transaction metadata, transaction graphs, and blockchain network data, pattern recognition algorithms can detect similarities between known fraud cases and ongoing fraudulent activities, enabling proactive detection and prevention of fraudulent transactions.

Furthermore, real-time monitoring and alerting systems can be implemented to detect and respond to fraudulent activities as they occur. These systems continuously monitor blockchain transactions, analyze transaction data in real time, and trigger alerts or notifications for suspicious activities. Automated response mechanisms, such as transaction freezing, blacklisting addresses, or triggering manual review processes, can be integrated into these systems to mitigate risks and prevent further fraudulent transactions [11].

Collaboration and information sharing among blockchain stakeholders, including exchanges, regulators, law enforcement agencies, and blockchain analytics firms, are also essential for detecting and combating fraudulent activities effectively. Sharing threat intelligence, fraud patterns, and suspicious transaction reports can help identify emerging threats, improve detection capabilities, and coordinate efforts to disrupt fraudulent schemes and prosecute perpetrators.

10.7.2 Anomaly detection techniques for blockchain networks

Anomaly detection techniques for blockchain networks have been explored in several papers. One approach is to use DL frameworks based on self-encoder and attention mechanisms, such as GraphAEAtt, which can extract high-dimensional features from the graph structure relationships [1]. Another approach is to utilize ML algorithms to detect blockchain attacks by training a federated learning-based anomaly detection system using aggregate data gathered from observing blockchain activity [2,6].

Anomaly detection techniques are essential for identifying unusual or suspicious behavior within blockchain networks, where transparency and immutability are critical but fraudulent activities can still occur.

Given the decentralized and distributed nature of blockchain technology, detecting anomalies requires specialized approaches tailored to the unique characteristics of blockchain networks. Here, we explore some key anomaly detection techniques used in blockchain networks.

Address Clustering Analysis: Address clustering involves grouping together blockchain addresses that are likely controlled by the same entity or user. By analyzing transaction patterns and network topology, address clustering techniques can identify clusters of addresses that exhibit similar behavior or are involved in common transactions. Anomalies may be detected when addresses deviate from their typical clustering patterns or exhibit unexpected behavior, such as sudden changes in transaction volume or connections to known illicit entities.

Transaction Graph Analysis: Transaction graph analysis leverages the inherent structure of blockchain data, representing transactions as nodes and addresses as edges in a graph. By analyzing the topology of the transaction graph, anomalies such as unusual transaction flows, hub addresses with high connectivity, or isolated clusters of addresses may be detected. Graph-based anomaly detection techniques, such as graph centrality measures or community detection algorithms, can uncover abnormal network structures or suspicious activity patterns within the blockchain network.

Temporal Analysis: Temporal analysis focuses on identifying anomalies based on changes in transaction patterns over time. By analyzing transaction timestamps, transaction frequency, or transaction volume trends, temporal anomaly detection techniques can detect sudden spikes or drops in activity, unusual patterns of transaction timing, or deviations from historical transaction behavior. These anomalies may indicate fraudulent activities, such as coordinated attacks or insider threats attempting to manipulate the blockchain network.

Consensus Mechanism Monitoring: Consensus mechanism monitoring involves analyzing the behavior of network participants and validators to detect anomalies in the consensus process. For example, in PoW blockchains, anomalies such as sudden changes in mining difficulty, hash rate fluctuations, or long forks in the blockchain may indicate attempted 51% attacks or mining pool manipulation. Similarly, in PoS blockchains, anomalies such as unusual voting patterns or stake concentration may signal attempts to compromise the integrity of the network.

Machine Learning-Based Approaches: Anomalies in blockchain networks can be detected by employing ML techniques, which encompass supervised and unsupervised learning algorithms. By utilizing labeled data, supervised learning models can be trained to classify transactions or network behavior as normal or anomalous based on predetermined criteria. However, unsupervised learning algorithms, such as clustering or autoencoders, can identify unusual patterns or outliers in the data without the need for labeled examples. By harnessing the power of ML, the process of anomaly detection in blockchain networks can be automated and adjusted to adapt to evolving threats and attack vectors.

10.7.3 Case studies and real-world applications

The promise of blockchain technology to transform existing processes, improve transparency, and stimulate innovation has garnered considerable interest from a range of businesses. The benefits and practical application of blockchain technology in a variety of industries are demonstrated by a large number of case studies and real-world applications.

Supply Chain Management: One of the most prominent uses of blockchain technology is in the management of supply chains. Prominent corporations such as Walmart and IBM have effectively integrated blockchain solutions to monitor the provenance and flow of goods from their point of origin to the final customer. Blockchain provides supply chain transparency, traceability, and authenticity by recording transactions on a distributed ledger. As a result, there are fewer instances of inefficiencies, fraud, and counterfeiting.

Financial Services: Because blockchain technology makes transactions faster, more secure, and more affordable, it has significantly altered the financial services sector. For example, Ripple uses blockchain to help financial institutions send money internationally, which cuts down on transaction fees and settlement times. In a similar vein, DeFi platforms like MakerDAO and Compound use blockchain technology to offer decentralized borrowing, trading, and lending services. This promotes financial inclusion for underprivileged groups by eschewing traditional middlemen.

Healthcare: Blockchain technology has the potential to totally alter the healthcare sector by ensuring interoperability, enhancing transparency in medical supply chains, and securely storing and transmitting patient data. Companies like Medicalchain and SimplyVital Health are actively developing blockchain-based solutions for pharmaceutical traceability, medical credentialing, and electronic health record (EHR) administration. These advancements provide individuals greater control over their health information while guaranteeing security and privacy.

Identity Management: Identity management systems can be greatly enhanced by the implementation of blockchain technology, as it provides a secure and tamper-proof record of individuals' identities and credentials. For example, Estonian people may now securely access government services, sign papers, and authenticate themselves online thanks to the government's effective integration of blockchain-based digital identification systems. In a similar vein, businesses like Civic and uPort are actively developing decentralized identity platforms that let consumers profit from and securely manage their personal information.

Chain of Supply Traceability: A further noteworthy use of blockchain technology is to guarantee the authenticity and traceability of goods in a variety of sectors. For instance, VeChain successfully uses blockchain to track and confirm the legitimacy of prescription drugs, food items, and luxury goods. This increases consumer confidence, lowers the availability of fake goods, and promotes sustainability by helping businesses monitor and lessen their environmental effect. It also allows consumers to confirm the provenance and quality of products.

10.8 BLOCKCHAIN ANALYTICS AND VISUALIZATION

10.8.1 Exploratory data analysis of blockchain networks

The examination and comprehension of the structure, behavior, and characteristics of blockchain data to gain insights and inform decision-making is an integral part of the Exploratory Data Analysis (EDA) of blockchain networks.

Investigative graph analysis of the Ethereum blockchain's network data was the main focus of Yap et al.'s study. To analyze transaction data, this research makes use of mathematical and statistical modeling in addition to network visualization. For EDA of blockchain networks, the special characteristics of blockchain technology—such as its immutability, transparency, and decentralization—present both potential and obstacles. We shall examine the essential elements and methods of EDA of blockchain networks in this paper [12].

The first step in the EDA of blockchain networks is the collection and preparation of data for analysis. Blockchain data is typically obtained from transaction records, block information, and network statistics obtained from blockchain explorers, APIs, or direct access to the blockchain network. Data preprocessing tasks may involve cleaning, filtering, and transforming the raw data to eliminate noise, handle missing values, and ensure consistency and integrity.

Descriptive Statistics: Descriptive statistics provide an overview of the basic characteristics and distribution of blockchain data. Key descriptive statistics include transaction counts, block sizes, transaction fees, transaction volumes, and time series analysis of network activity over time. Descriptive statistics help identify trends, patterns, and anomalies in blockchain data and provide insights into network dynamics and behavior.

Transaction Analysis: Transaction analysis focuses on understanding the flow and behavior of transactions within the blockchain network. This involves examining transaction attributes such as sender and receiver addresses, transaction amounts, transaction types, and timestamps. Transaction analysis techniques include visualizations such as histograms, pie charts, and scatter plots to explore transaction patterns, identify outliers, and detect suspicious activities.

Network Topology Analysis: Network topology analysis involves studying the structure and connectivity of the blockchain network. This includes analyzing the distribution of nodes, the degree distribution of the network, and the connectivity between nodes. Network analysis techniques such as graph theory, centrality measures, and community detection algorithms help identify key nodes, clusters, and subnetworks within the blockchain network.

Temporal Analysis: Temporal analysis focuses on understanding how blockchain data evolves over time. This includes analyzing transaction timestamps, block intervals, and network activity trends. Temporal analysis techniques such as time series analysis, trend analysis, and seasonality detection help identify patterns, cyclicality, and anomalies in blockchain data over different time periods.

Blockchain addresses that are most likely under the control of the same entity or person are grouped together using address clustering techniques. This makes it easier to identify groups of addresses linked to particular businesses, such as wallets, exchanges, or people. Within the blockchain network, address clustering makes it easier to identify entities, attribute transactions, and analyze user behavior [10,13,14].

The goal of anomaly detection techniques is to locate odd or questionable activity happening throughout the blockchain network. This entails looking for anomalies, departures from the norm, and possible security risks like 51% or double-spending attacks. Statistical analysis, ML algorithms, and heuristics-based techniques customized for blockchain data properties are some examples of anomaly detection techniques.

10.8.2 Visualizing blockchain data for insights

Visualizing blockchain data is a powerful technique for gaining insights, understanding patterns, and communicating complex information in a more intuitive and accessible manner. Vinceslas et al. [8] proposed an abstraction layer architecture to improve the auditability and intuitiveness of complex business analysis of distributed ledger systems. By utilizing various visualization techniques, stakeholders can effectively explore and analyze blockchain data, uncovering valuable insights and informing decision-making processes. In this paper, we will explore the importance of visualizing blockchain data and discuss key visualization techniques used in this context.

Understanding Complex Relationships: Blockchain data often involves complex relationships between transactions, addresses, blocks, and network participants. Visualization techniques such as network graphs, Sankey diagrams, and chord diagrams help illustrate these relationships visually, enabling stakeholders to understand the flow of transactions, identify clusters of addresses, and explore the connectivity and interdependencies within the blockchain network.

Identifying Transaction Patterns: Visualizing transaction patterns is essential for understanding the behavior of users, identifying trends, and detecting anomalies within the blockchain network. Time series plots, heatmaps, and histograms can be used to visualize transaction volumes, transaction frequencies, and transaction values over time. By visualizing transaction patterns, stakeholders can identify peak periods of activity, detect irregularities, and assess the overall health of the blockchain network.

Exploring Network Topology: Visualizing the network topology of the blockchain network helps stakeholders understand the structure, connectivity, and centrality of nodes within the network. Techniques such as node-link diagrams, force-directed layouts, and tree maps can be used to visualize the distribution of nodes, the degree distribution of the network, and the relationships between nodes. Network topology visualizations enable stakeholders to identify key nodes, detect clusters, and assess the resilience and robustness of the blockchain network [14].

Tracking Transaction Flows: Visualizing transaction flows within the blockchain network provides insights into the movement of assets, the path of transactions, and the distribution of funds across addresses. Flow diagrams, Sankey diagrams, and flowcharts can be used to visualize transaction flows, highlighting the origins, destinations, and intermediaries involved in transactions. By tracking transaction flows visually, stakeholders can detect suspicious activities, trace the movement of funds, and assess the efficiency of transaction processing within the blockchain network.

Monitoring Market Dynamics: Visualizing market dynamics within blockchain networks, such as cryptocurrency markets, helps stakeholders analyze price movements, trading volumes, and market trends. Candlestick charts, line charts, and scatter plots can be used to visualize price fluctuations, trading volumes, and market indicators over time. By monitoring market dynamics visually, stakeholders can identify trading patterns, assess market sentiment, and make informed decisions in cryptocurrency trading and investment.

Detecting Anomalies and Security Threats: Visualizing anomalies and security threats within the blockchain network helps stakeholders identify irregularities, detect potential attacks, and mitigate risks effectively. Anomaly detection visualizations, such as heatmaps, scatter plots, and box plots, can be used to visualize outliers, deviations from expected behavior, and potential security breaches within the blockchain network. By visualizing anomalies and security threats, stakeholders can take proactive measures to safeguard the integrity and security of the blockchain ecosystem.

10.9 PRIVACY PRESERVATION AND DE-ANONYMIZATION TECHNIQUES

10.9.1 Preserving user privacy in blockchain transactions

Because blockchain technology is inherently transparent and immutable, protecting user privacy in transactions is very important. Although blockchain technology has many benefits, like data integrity and decentralization, it also has drawbacks with regard to user confidentiality and privacy. To protect user privacy in blockchain transactions, a number of strategies and tactics are used in this context.

Using cryptographic techniques like encryption, ZKPs, and ring signatures is one of the main ways to keep users' privacy safe in blockchain transactions. Techniques for encrypting data make sure that private information, including transaction amounts or user identities, is hidden from prying eyes before being added to the blockchain. With ZKPs, one party (the prover) can show the other party (the verifier) that a statement is true without revealing any further information. As a result, consumers can transact without disclosing any personal information. Users can sign transactions on behalf of a group using ring signatures, which makes it difficult to identify which group member began the transaction, thereby ensuring anonymity.

Preserving user privacy in blockchain transactions can also be achieved by using blockchain platforms and protocols that give privacy priority. These platforms—Monero, Zcash, and Dash, for example—are made with the express purpose of protecting user privacy through the deployment of cutting-edge cryptographic methods like stealth addresses, ring signatures, and zk-SNARKs (Zero-Knowledge Succinct Non-Interactive Arguments of Knowledge). By ensuring that transaction details, including sender and recipient addresses and transaction amounts, are hidden or masked, these measures protect user privacy while upholding the blockchain's integrity and security.

To improve user privacy in blockchain transactions, privacy-enhancing technology like mixers and tumblers is utilized in addition to cryptographic methods and privacy-focused blockchain platforms. In order to make it difficult to track down individual transactions, mixers and tumblers combine several transactions into one and blend them together. By doing this, people taking part in blockchain transactions can benefit from increased privacy and anonymity as the link between sender and recipient addresses is effectively severed [15].

Furthermore, the maintenance of user privacy in blockchain transactions is greatly dependent on regulatory compliance and data protection procedures. The handling and processing of personal data, including data held on blockchain networks, is subject to strict regulations, such as the General Data Protection Regulation (GDPR) in the European Union. To guarantee that user privacy rights are respected and protected, blockchain developers and operators must comply with these requirements by putting in place privacy-enhancing measures such as data minimization, pseudonymization, and user consent methods.

10.9.2 De-anonymization attacks and countermeasures

De-anonymization attacks pose a significant threat to user privacy in blockchain transactions, as they aim to reveal the identities of users participating in transactions on the blockchain. These attacks exploit vulnerabilities in blockchain networks and privacy-enhancing technologies to uncover sensitive information about users, such as their real-world identities, transaction history, and financial activities. In response to these threats, various countermeasures and mitigation strategies have been developed to enhance user privacy and protect against de-anonymization attacks.

One common de-anonymization attack is known as network analysis, where adversaries analyze the structure and connectivity of the blockchain network to infer relationships between users and trace transactions back to their origin. By analyzing transaction patterns, network topology, and transaction flows, attackers can identify clusters of addresses associated with specific users or entities, enabling them to de-anonymize users and uncover sensitive information.

Another de-anonymization technique is known as de-anonymization through auxiliary information. This attack leverages external sources of information, such as social media profiles, online forums, or leaked databases, to link blockchain addresses to real-world identities. By correlating blockchain addresses with publicly available information, attackers can identify individuals participating in blockchain transactions and potentially expose their private activities.

Sybil attacks are another issue that raises questions about user privacy in blockchain networks. Attackers use several false identities or pseudonymous accounts in these types of attacks in an attempt to take over a sizable chunk of the network. Attackers can alter the network's consensus, control the flow of transactions, and conduct targeted assaults against certain users or transactions by controlling a sizable number of network nodes or addresses.

To address de-anonymization attacks and safeguard the privacy of users in blockchain transactions, various countermeasures and strategies have been devised:

1. Use of Privacy-Enhancing Technologies: To improve user anonymity, privacy-focused blockchain platforms and protocols, like Monero, Zcash, and Dash, integrate cutting-edge cryptography approaches and privacy features. These platforms conceal transaction data and

protect user privacy by using techniques like ring signatures, stealth addresses, and ZKPs.

2. Utilizing CoinJoin and Mixing Services: By combining their transactions with those of other users, users can use CoinJoin and mixing services to make it more difficult for adversaries to track down the source of a transaction. These services combine transactions from several users, resulting in a jumbled transaction history that hides the relationship between the source and recipient addresses [16].
3. Prevention of Address Reuse: Avoiding the reuse of addresses can help mitigate the risk of de-anonymization attacks. Reusing addresses can facilitate the linking of multiple transactions to the same user, making it easier for attackers to identify and exploit user information. By using new addresses for each transaction or implementing hierarchical deterministic (HD) wallets, the likelihood of address reuse is reduced, thereby enhancing user privacy.
4. Enhancements in Network and Transactional Privacy: Implementing privacy enhancements at the network level, such as onion routing or mixnets, can obscure transaction metadata and prevent adversaries from monitoring the flow of transactions. Furthermore, adopting transactional privacy enhancements like confidential transactions or bulletproofs can conceal transaction amounts and further elevate user privacy.

Apart from the technological solutions outlined above, regulatory compliance and following best practices are essential for protecting user privacy and reducing the likelihood of de-anonymization assaults. Privacy protection can be enhanced by adhering to legal standards, such as those pertaining to data protection and anti-money laundering (AML) legislation. Adopting best practices for user consent, pseudonymization, and data minimization can help improve privacy protection and lower the risk of user data exposure.

10.9.3 Privacy-preserving machine learning techniques for blockchain

Privacy-preserving ML techniques have a significant role in safeguarding sensitive data while extracting valuable insights from blockchain networks. The increasing adoption of blockchain technology in various industries emphasizes the importance of maintaining data privacy and confidentiality. In this regard, privacy-preserving ML techniques provide innovative solutions to address privacy concerns while leveraging the advantages offered by blockchain technology. This article explores several key privacy-preserving ML techniques applicable to blockchain applications.

A cryptographic technique called homomorphic encryption makes it possible to do calculations on encrypted material without requiring its decryption. This ensures the secrecy of sensitive data and allows ML algorithms that protect privacy to analyze encrypted data stored on the blockchain. Homotopy encryption efficiently safeguards user confidentiality and privacy by encrypting data before it is processed or stored on the blockchain. Predictive analytics and data sharing are two areas where it finds use in industries like healthcare, finance, and other delicate disciplines.

Secure Multi-Party Computation (MPC) is a cryptographic protocol that enables multiple parties to jointly compute a function over their inputs while maintaining the privacy of those inputs. In the context of blockchain, MPC allows participants to collaborate on ML tasks, such as model training or prediction, without revealing their private data to each other or third parties. This ensures privacy and confidentiality while facilitating collaborative ML in decentralized environments.

Differential privacy is a privacy-preserving technique that modifies the results of sensitive data queries or computations in order to introduce noise and hence offer strong privacy assurances. Differential privacy allows for accurate aggregate analysis while preventing adversaries from deriving sensitive information about individual data points by adding regulated levels of noise. ML models based on blockchain data are protected against unauthorized disclosure of sensitive information by using differential privacy strategies.

Federated Learning is a decentralized ML technique in which only model updates are exchanged with a central server or aggregator, while model training is done locally on edge devices or nodes. This eliminates the need to share raw data and allows collaborative model training across remote data sources, like blockchain nodes. Federated learning preserves user privacy by storing sensitive data locally on devices and facilitating central sharing and aggregation of model advances. It is especially appropriate for blockchain network applications when privacy is a concern.

ZKPs are cryptographic procedures that let the prover (one party) show the verifier (another party) that a statement is true without revealing any further information. ZKPs can be utilized inside the blockchain framework to confirm the precision of calculations or transactions without concealing confidential inputs or information. By doing this, the blockchain's transaction authenticity and integrity are guaranteed, and user privacy is protected.

10.10 NETWORK SECURITY AND ATTACK DETECTION

10.10.1 Understanding the security of blockchain networks

Gaining an understanding of blockchain security is essential to understanding the strength and dependability of blockchain technology, which forms the basis for a wide range of industries and applications. Blockchain, which was first created as the foundational technology for Bitcoin, has matured into a flexible framework with a wide range of uses, including supply chain management and banking. Its distributed, decentralized architecture along with cryptography concepts give it intrinsic security. But in order to truly understand blockchain network security, one needs to examine all of its essential elements, possible weak points, and security risk-mitigation techniques.

The potential of blockchain technology in network security has attracted attention. It provides auditability, security, anonymity, immutability, and decentralization. Blockchain overcomes the restrictions and difficulties related to its application in network security by offering a distributed, transparent, and impenetrable ledger for safe data movement throughout the network. A sociotechnical security analysis of blockchain systems reveals discrepancies between the social, technical, and infrastructural layers that impact the security and trust assumptions. Moreover, blockchain has been the subject of investigation in the field of cybersecurity, where it has proven beneficial in addressing security issues and offering advantages such as data integrity and confidentiality.

Decentralization, consensus techniques, cryptography, an immutable ledger, and network design are all parts of blockchain network security. Blockchain networks are guaranteed to function via a dispersed network of nodes rather than depending on a central authority thanks to decentralization. By removing single points of failure and lowering the possibility of hostile attacks or manipulation, this improves security. Consensus systems, such as DPoS, PoW, PoS, and Practical Byzantine Fault Tolerance (PBFT), guarantee that network users agree on the legitimacy of transactions and the sequence in which they should be recorded on the blockchain. By preventing double-spending, these techniques protect the blockchain network's integrity.

Blockchain network security is greatly aided by cryptography, which offers digital signature, encryption, and authentication techniques. Public-key cryptography is a widely used technique for generating cryptographic key pairs, signing transactions, confirming asset ownership, and enabling safe communication between network users. A transaction cannot be changed or removed once it has been recorded because to the blockchain ledger's immutability. It is computationally impossible to change previous transactions without the consent of the majority of network users, a feature made possible by cryptographic hash functions and the consensus process. Peer-to-peer (P2P) networking is used in blockchain networks, where nodes interact with one another directly and without the need for middlemen. By facilitating redundancy, consensus, and data distribution, this architecture improves the blockchain network's resilience and security. While blockchain technology presents robust security features, it is not impervious to vulnerabilities and threats. Blockchain networks face a number of common risks and vulnerabilities, such as the 51% attacks. A 51% attack occurs in PoW consensus systems when a single person or organization has a disproportionate amount of the network's hashing power. This gives the attacker the ability to alter transactions, undo transactions, or spend coins twice.

One such weakness is the Sybil attacks. In Sybil attacks, several fictitious identities or nodes are created in an attempt to take control of the network or sway the consensus process. Because these attacks interfere with consensus processes and spread misleading information, they jeopardize the security and integrity of the blockchain network.

Vulnerabilities in smart contracts present an additional danger. Self-executing contracts with predefined circumstances, or "smart contracts," can be exploited and have security holes or coding faults. Smart contract vulnerabilities put users and blockchain applications at risk of theft, unauthorized activity, or money loss.

Furthermore, privacy risks emerge despite the pseudonymous nature of blockchain transactions. The traceability of transactions on the public ledger allows for the analysis of transaction patterns, network traffic, and metadata, which can unveil sensitive information about users and their transactions, posing a threat to privacy.

Double-spending is also a concern. It occurs when a user spends the same digital asset multiple times, leading to inconsistencies in the blockchain ledger. While consensus mechanisms such as PoW and PoS prevent double-spending, the risk persists in permissionless blockchain networks.

To mitigate security risks and enhance blockchain network security, several strategies and best practices are employed. Diversifying consensus mechanisms through hybrid consensus models reduces the risk of 51% attacks and enhances network security. Transparent governance mechanisms ensure that decisions regarding network upgrades, protocol changes, and security enhancements are made collaboratively and transparently. Regular code audits and security assessments of blockchain protocols, smart contracts, and network infrastructure help identify and mitigate vulnerabilities early in the development lifecycle.

On the blockchain, user privacy and confidentiality are improved by implementing privacy-enhancing technologies like homomorphic encryption, ring signatures, and ZKPs. Finally, the maintenance of a safe and robust blockchain ecosystem depends on educating developers, users, and stakeholders on blockchain security best practices, dangers, and mitigation techniques. Workshops, security awareness campaigns, and training sessions educate participants about security risks and provide them with the tools they need to take preventative action to safeguard the network and themselves.

10.10.2 Detecting and mitigating attacks with machine learning

ML has become a powerful tool for strengthening cybersecurity defenses across a range of disciplines when used in attack detection and mitigation. It is frequently insufficient to defend against emerging threats to rely exclusively on traditional security measures, given the growth of digital threats and the increasing sophistication of assaults. Using ML techniques in this situation can help cybersecurity measures become more resilient and effective by quickly identifying, analyzing, and reacting to cyber threats in real time. This talk explores how ML can be used to detect and lessen assaults, as well as the difficulties that come with using it and best practices for putting ML-based security solutions into practice.

10.10.2.1 The application of machine learning in cybersecurity

Without the need for explicit programming, ML techniques help computer systems learn from data, identify patterns, and make predictions or judgments. In cybersecurity, ML techniques are used to analyze large amounts of security data, including system logs, network traffic, and user behavior. The goal of this analysis is to find abnormalities, identify malicious activity, and mitigate cyber threats. Security systems based on ML have many benefits, such as automation, scalability, and adaptability to changing threats.

10.10.2.2 Detection of cyberattacks through machine learning

Anomaly Detection: This popular ML technique is used to find odd or suspicious activity in cybersecurity data. ML algorithms acquire knowledge of the standard patterns and actions displayed by networks, systems, and users. They then identify any departures from these standard practices as possible anomalies. Methods including density estimation, clustering, and unsupervised learning are used to find patterns in system logs, network traffic, and user activity. Signature-based Detection: This method of identifying known threats or malicious activity uses signatures, which are preset rules or patterns. By using previous data to identify these characteristics, ML algorithms may be trained to categorize new instances based on how similar they are to existing dangers. It has been shown that signature-based detection works effectively for identifying viruses, malware, and other common attack vectors.

Behavioral analysis is the process of keeping an eye on and assessing how individuals, devices, and apps behave in order to spot any abnormalities that might point to malicious activity. Supervised learning and reinforcement learning are two ML techniques that can be used to learn behavior profiles and quickly spot any abnormal behavior. When it comes to identifying insider threats, advanced persistent threats (APTs), and zero-day assaults, behavioral analysis is especially useful.

Threat Intelligence Integration: To improve their ability to detect threats, ML algorithms can take advantage of threat intelligence feeds, vulnerability databases, and security advisories. The identification of zero-day vulnerabilities, new threats, and indicators of compromise (IOCs) can be enhanced by ML-based security solutions through the integration of external threat intelligence sources with internal security data.

10.10.2.3 Mitigating cyberattacks with machine learning

i. ML-based security systems automate incident response, allowing for real-time identification and mitigation of cyber threats. Automated response mechanisms, such as adaptive access restrictions, network segmentation, and dynamic policy enforcement, may be implemented using ML for threat analysis and risk assessment.
ii. Predictive Maintenance: ML algorithms can forecast security events by analyzing past data, finding patterns of vulnerabilities, and proactively applying preventative measures. Predictive maintenance approaches help organizations anticipate and address security problems before they become full-fledged assaults.
iii. Adaptive Defense Strategies: ML-based security solutions enable adaptive defense strategies that dynamically adjust security measures based on evolving threat landscapes and changing risk profiles. Adaptive defense mechanisms, such as dynamic threat modeling, adaptive authentication, and self-learning security controls, continuously adapt to emerging threats and mitigate risks in real time.
iv. Threat Hunting and Attribution: ML algorithms can assist security analysts in threat hunting and attribution by correlating disparate security data, identifying attack patterns, and attributing malicious activities to specific threat actors or groups. ML-driven threat hunting techniques, such as behavior-based clustering, similarity analysis, and link analysis, enhance the effectiveness of cybersecurity investigations and incident response efforts.

10.10.2.4 Challenges and considerations

While ML-based security solutions provide major benefits, there are several issues and concerns that must be addressed:

i. Data Quality and Labeling: To learn successfully, ML systems require high-quality training data that has been labeled. Ensuring the quality, completeness, and relevance of training data is critical for developing strong ML models for cybersecurity.
ii. Adversarial Attacks: Adversarial attacks aim to deceive ML models by manipulating input data to produce incorrect outputs. Adversarial robustness techniques, such as adversarial training, model ensembling, and input perturbation, are essential for defending against adversarial attacks in cybersecurity.
iii. Model Interpretability: Comprehending and explaining the decisions made by ML models are crucial for fostering trust, comprehending behavior, and preventing prejudice or discrimination. Model interpretability approaches, including feature significance analysis, model visualization, and explanation generation, improve the transparency and accountability of ML-based security solutions.

iv. Scalability and Performance: ML-based security systems require scalability and performance to manage massive amounts of data, real-time processing, and many attack scenarios. In cybersecurity, solutions for boosting scalability and performance include optimizing ML algorithms, establishing distributed computer infrastructure, and using cloud services.

10.10.2.5 Best practices for implementing ML-based security solutions

i. Continuously monitor and assess ML models for performance, accuracy, and efficacy. Regular updates, retraining, and validation are critical for sustaining the relevance and reliability of ML-based security solutions over time.
ii. Collaborative Threat Intelligence Sharing: Collaborative threat intelligence sharing among organizations, industries, and cybersecurity communities enhances collective defense against cyber threats. Sharing threat data, IOCs, and actionable intelligence enables timely detection and mitigation of cyberattacks.
iii. Human-Machine Collaboration: Human-machine collaboration is essential for effective cybersecurity operations. ML-based security solutions should augment, rather than replace, human expertise and decision-making. Security analysts play a critical role in validating ML-generated alerts, investigating security incidents, and making informed decisions based on ML-driven insights.
iv. Privacy and Ethical Considerations: Protecting user privacy, respecting ethical principles, and adhering to regulatory requirements

10.10.3 Enhancing resilience against network attacks

Enhancing resilience against network attacks is crucial in today's digital landscape, where cyber threats are continuously evolving and becoming more sophisticated. Network attacks pose significant risks to organizations, governments, and individuals, leading to data breaches, financial losses, and reputational damage. To combat these threats effectively, it is essential to implement robust cybersecurity measures that enhance the resilience of network infrastructure and mitigate the impact of attacks. This article explores strategies for enhancing resilience against network attacks, including proactive defense measures, incident response strategies, and best practices for building resilient network architectures.

10.10.3.1 Understanding network attacks

A vast array of malevolent actions with the intent to jeopardize the availability, confidentiality, or integrity of network resources are collectively referred to as network assaults. Typical kinds of cyberattacks consist of the following:

Attacks known as denial of service (DoS) and distributed denial of service (DDoS) overload servers and routers with excessive traffic, blocking them from being used by authorized users. Malware and ransomware attacks: To steal data, interfere with operations, or demand ransom payments, malicious software—such as viruses, worms, and ransomware—infiltrates network systems. Phishing and social engineering attacks: Users are tricked into disclosing private information, including login passwords or financial information, by means of phishing emails, phony websites, and social engineering techniques.

10.10.3.1.1 Strategies for enhancing resilience against network attacks

1. The application of Protection-in-A security strategy called "depth" uses a variety of security controls at multiple network architectural layers. These controls include internal ones like network segmentation and endpoint security solutions, as well as perimeter defenses like firewalls and intrusion detection system (IDS). Organizations can identify and stop network assaults at different points in the cyber death chain by implementing numerous levels of defense.
2. In order to recognize and react to network threats instantly, ongoing monitoring and threat detection are essential. System logs, network traffic, and user behavior must all be regularly observed. The detection of suspicious activity and possible security breaches can be aided by anomaly detection algorithms, security information and event management (SIEM) systems, and intrusion detection and prevention system (IDPS). Through the analysis of massive amounts of security data and the identification of patterns suggestive of harmful behavior, the application of ML and artificial intelligence techniques can further improve threat detection capabilities.
3. Hardening network infrastructure involves the implementation of security best practices and configuration settings to reduce the attack surface and minimize vulnerabilities. This includes regular patching and updating of software, disabling unnecessary services and protocols, and implementing strong authentication mechanisms like multi-factor authentication (MFA) and encryption.
4. Using configuration settings and security best practices to lower attack surfaces and minimize vulnerabilities is known as hardening network infrastructure. This entails applying software patches and updates on a regular basis, turning down unused services and protocols, and putting robust authentication measures like encryption and MFA in place.
5. Creating and testing an incident response plan on a regular basis is essential to responding to network threats efficiently and reducing their effects. In the case of a security incident, the incident response plan should specify roles and responsibilities, escalation procedures, communication protocols, and recovery actions. By running

simulations and tabletop exercises, staff members are guaranteed to be ready to react swiftly and efficiently to cyberattacks [7].

6. Creating a security-aware culture inside an organization requires educating staff members on social engineering techniques, threat awareness, and cybersecurity best practices. Programs for security awareness training ought to address subjects including safe browsing practices, password hygiene, and phishing awareness. Establishing employee empowerment to identify and communicate security concerns can help organizations improve their resistance to network intrusions.
7. Collaboration and information sharing among organizations, industry sectors, and cybersecurity communities are crucial for collectively defending against network attacks. The sharing of threat intelligence, attack indicators, and best practices enables organizations to identify emerging threats, respond proactively, and enhance their overall resilience against cyber threats.

10.10.3.2 Building resilient network architectures

i. Building resilient network architectures necessitates an all-encompassing approach that amalgamates technology, processes, and individuals. Fundamental principles for constructing resilient network architectures incorporate:
ii. Redundancy and Failover Mechanisms: The implementation of duplicated network components, such as backup servers, redundant power supplies, and failover mechanisms, guarantees heightened availability and mitigates downtime in the event of a network failure or attack.
iii. Scalability and Elasticity: The design of network architectures that are scalable and elastic equips organizations to accommodate escalating traffic demands and adapt to evolving business requirements. Cloud-based solutions and virtualized network functions foster scalability and agility in response to developing threats.
iv. Automation and Orchestration: Reducing the possibility of human mistake and increasing operational efficiency are two benefits of automating standard network administration tasks including provisioning, configuration, and monitoring. Organizations can mitigate the effects of network assaults and respond quickly to security problems by utilizing orchestration technologies and automation frameworks.
v. Communication Protocols That Are Resilient: Resilient communication protocols, such as IPsec, Secure Shell (SSH), and Transport Layer Security (TLS), are used to prevent unauthorized access to network conversations and to protect data while it is in transit. Data communicated across a network is guaranteed to be authentic, confidential, and integrity thanks to secure communication methods.

vi. Continuous Evaluation and Improvement: Continuously assessing the effectiveness of network security measures and adapting to emerging threats is indispensable for upholding resilience against network attacks. Regular security assessments, penetration testing, and risk assessments aid in identifying vulnerabilities in the network architecture and accordingly prioritizing security investments.

10.11 CHALLENGES, FUTURE DIRECTIONS, AND EMERGING TRENDS

i. A number of difficulties and roadblocks need to be addressed in order to fully realize the promise of blockchain technology in DL systems when examining the convergence of ML and DL with blockchain technology for intelligence applications. The following are included in the final thoughts and important suggestions:
ii. Monitoring the volume and type of data used to train DL models through the use of blockchain's data traceability, immutability, and integrity features has a lot of potential. But data quality problems are hard for existing blockchain systems to handle efficiently, especially in delicate industries like transportation and healthcare.
iii. The efficacy of current blockchain-based DL systems is significantly impacted by key performance indicators, such as system throughput, execution delay, block propagation time, data volume, competing interests among participants, and smart contract vulnerabilities.
iv. While public blockchain platforms are vulnerable to data privacy breaches because of their zero-access control policy, private blockchain platforms provide data privacy through private channels and access control policies. However, public blockchain systems are superior at recording the development of DL models at every step of their construction, modification, or application.
v. The scale of the blockchain network has a significant impact on the scalability of blockchain-based applications. These applications can function more efficiently when DL techniques are used for data compression and redundant data minimization, especially as networks get larger.

10.11.1 Future directions and emerging trends

1. The future direction and emerging trends in the convergence of blockchain intelligence and ML and DL have the potential to have a significant impact on intelligence applications across a variety of industries, building upon the challenges and suggestions previously discussed. We outline some possible directions for further study and invention in this section.

2. Improved Data Quality Control: Future research endeavors ought to focus principally on creating resilient methods that guarantee data quality in blockchain-based DL systems. To address issues with data correctness, completeness, and consistency, this include investigating methods for data validation, verification, and cleansing. Furthermore, developments in lineage tracing and data provenance can offer insightful information about the beginnings and development of data within blockchain networks.
3. Optimized Performance Metrics: More investigation is needed to improve system throughput, execution latency, and block propagation time in particular when it comes to performance metrics in blockchain-based DL systems. New consensus techniques like sharding and PoS could increase efficiency and scalability. Computational performance may also be improved by developments in hardware acceleration technologies, such as quantum computing or specialized DL processors.
4. Privacy-Preserving Techniques: Future advancements in privacy-preserving methods for blockchain-based DL systems are eagerly awaited, considering the importance of data privacy. This involves investigating cutting-edge cryptographic techniques like homomorphic encryption and secure MPC to allow for private and secure data sharing and calculation without sacrificing confidentiality. Additionally, while preserving the integrity of blockchain transactions, user privacy can be improved by integrating differential privacy methods and ZKPs.
5. Scalability Solutions: Scalability remains a significant challenge for blockchain networks, particularly as the size and complexity of DL models and datasets continue to increase. Research efforts should focus on developing scalable solutions for DL systems based on blockchain, including off-chain computation, layer-two scaling solutions (e.g., state channels and sidechains), and network optimization techniques. Moreover, advancements in interoperability protocols and cross-chain communication mechanisms can facilitate seamless data exchange and collaboration across diverse blockchain networks.
6. Interdisciplinary Collaboration: Experts in blockchain technology, ML, cryptography, and domain-specific domain-specific areas should collaborate interdisciplinary on future research projects. Through fostering interdisciplinary collaborations, scholars can capitalize on a range of viewpoints and proficiencies to tackle intricate problems and stimulate novelty in DL applications grounded on blockchain technology. Research consortia, hackathons, and open-source communities are examples of collaborative projects that can promote information exchange and quicken scientific advancement.
7. Legal and Ethical Issues: As blockchain-based DL systems develop further, legal and ethical issues pertaining to data security, privacy, and responsibility must be addressed. Subsequent investigations ought

to focus on formulating standards, guidelines, and regulatory frameworks to guarantee the conscientious implementation and utilization of blockchain technology in intelligence applications. Initiatives that support accountability, justice, and openness in algorithmic decision-making can also assist reduce risks and guarantee the moral application of blockchain-based DL systems.

In conclusion, there is a great deal of potential for revolutionizing intelligence applications across industries in the future directions and developing trends in the integration of ML/DL with blockchain intelligence. Researchers and practitioners may fully realize the potential of blockchain-based DL systems and propel significant improvements in automation, decision-making, and intelligence analysis by tackling important issues, welcoming innovation, and promoting multidisciplinary collaboration.

REFERENCES

1. Shafay, M., Ahmad, R. W., Salah, K., Yaqoob, I., Jayaraman, R., & Omar, M. (2023). Blockchain for deep learning: review and open challenges. *Cluster Computing* 26(1), 197–221.
2. Uddin, M., Selvarajan, S., Obaidat, M., UlArfeen, S., Khadidos, A. O., Khadidos, A. O., & Abdelhaq, M. (2023). From hype to reality: unveiling the promises, challenges and opportunities of blockchain in supply chain systems. *Sustainability* 15(16), 12193. https://doi.org/10.3390/su151612193.
3. Li, Z., Sun, W., Zhan, D., Kang, Y., Chen, L., Bozzon, A., & Hai, R. (2024). Amalur: data integration meets machine learning. *IEEE Transactions on Knowledge and Data Engineering*, Anaheim, CA, USA.
4. Selvarajan, S. & Mouratidis, H. (2023). A quantum trust and consultative transaction-based blockchain cybersecurity model for healthcare systems. *Scientific Reports* 13(1), 7107. doi: 10.1038/s41598-023-34354-x.
5. Manoharan, H., Manoharan, A., Selvarajan, S., & Venkatachalam, K. (2023). Implementation of internet of things with blockchain using machine learning algorithm: enhancement of security with blockchain. In Najar, T. et al. (eds.), *Handbook of Research on Blockchain Technology and the Digitalization of the Supply Chain.* IGI Global: Hershey, PA, pp. 399–430. https://doi.org/10.4018/978-1-6684-7455-6.ch019.
6. Shitharth, S., Manoharan, H., Shankar, A., Alsowail, R. A., Pandiaraj, S., Edalatpanah, S. A., & Viriyasitavat, W. (2023). Federated learning optimization: a computational blockchain process with offloading analysis to enhance security. *Egyptian Informatics Journal* 24(4), 100406. https://doi.org/10.1016/j.eij.2023.100406.
7. Aluvalu, R., Kumaran, V. N. S., Thirumalaisamy, M., Basheer, S., Ali aldhahri, E., & Selvarajan, S. (2023). Efficient data transmission on wireless communication through a privacy-enhanced blockchain process. *PeerJ Computer Science.* https://doi.org/10.7717/peerj-cs.1308.

8. Vinceslas, L., Dogan, S., Sundareshwar, S., & Kondoz, A. M. (2023). Abstracting data in distributed ledger systems for higher level analytics and visualizations. *Future Internet* 15(1), 33.
9. Selvarajan, S., Srivastava, G., Khadidos, A. O., Khadidos, A. O., Baza, M., Alsheri, A., Lin, J. C.-W. (2023). An artificial intelligence lightweight blockchain security model for security and privacy in IIoT systems. *Journal of Cloud Computing* 12, 38. https://doi.org/10.1186/s13677-023-00412-y.
10. Diwan, P., Khandelwal, B., Dewangan, B. K., & Shriwas, P. (2023). A bibliometric analysis of network security on blockchain. In *2022 OPJU International Technology Conference on Emerging Technologies for Sustainable Development (OTCON)*, pp. 1–6. IEEE, Raigarh, Chhattisgarh, India.
11. Selvarajan, S., Manoharan, H., Iwendi, C., Al-Shehari, T., Al-Razgan, M., & Alfakih, T. (2023). SCBC: smart city monitoring with blockchain using Internet of Things for and neuro fuzzy procedures. *Mathematical Biosciences and Engineering* 20(12), 20828–20851. doi: 10.3934/mbe.2023922.
12. Natarajan, K., Karthikeyan, R., & Rajalingam, S. (2023). Importance of drone technology in agriculture. In Mohanty, S. N., Ravindra, J. V. R., Surya Narayana, G., Pattnaik, C. R., & Mohamed Sirajudeen, Y. (eds.), *Drone Technology*. Wiley: Hoboken, NJ, pp. 351–374. https://doi.org/10.1002/9781394168002.ch14.
13. Karthikeyan, R., Sundaravadivazhagan, B., Cyriac, R., Balachandran, P. K., & Shitharth, S. (2023). Preserving resource handiness and exigency-based migration algorithm (PRH-EM) for energy efficient federated cloud management systems. Mobile Information Systems 11. https://doi.org/10.1155/2023/7754765.
14. Srinivasarao, U., Karthikeyan, R., Sarangi, P. K., & Panigrahi, B. S. (2022). Enhanced movie recommendation and sentiment analysis model achieved by similarity method through Cosine and Jaccard similarity algorithms. *2022 International Conference on Computing, Communication, and Intelligent Systems (ICCCIS)*, Greater Noida, India, pp. 214–218. doi: 10.1109/ICCCIS56430.2022.10037722.
15. Hussain, S. F. M., Karthikeyan, R., Ramamoorthi, S., Arafat, I. S., & Gani, S. S. M. (2023). Enhanced protection for information and network using intrusion detection system. *2023 2nd International Conference on Edge Computing and Applications (ICECAA)*, Namakkal, India, pp. 280–286. doi: 10.1109/ICECAA58104.2023.10212144.
16. Saleem Raja, A., Balasubaramanian, S., Ganesan, P., Rajasekaran, J., & Karthikeyan, R. (2023). Weighted ensemble classifier for malicious link detection using natural language processing. *International Journal of Pervasive Computing and Communications*. https://doi.org/10.1108/IJPCC-09-2022-0312.

Chapter 11

HSMAO

Efficient resource allocation in cloud-fog Internet of Things (IoT) networks using metaheuristic scheduling algorithm

Santhosh Kumar Medishetti, VNLN Murthy, Ruqqaiya Begum, and Ganesh Reddy Karri

11.1 INTRODUCTION

Cloud-Fog Computing (CFC) has revolutionized the landscape of distributed computing by offering a flexible and scalable infrastructure for hosting diverse applications and services [1]. This paradigm leverages both centralized cloud data centers and distributed fog nodes located closer to end-users to efficiently process and manage computing tasks. Task Scheduling (TS) plays a pivotal role in optimizing resource utilization, minimizing latency, and enhancing overall system performance in CFC environments.

Traditional TS approaches often face challenges in dynamically allocating resources to meet the varying demands of applications and users. The inherent complexity, heterogeneity, and dynamic nature of CFC environments require innovative optimization techniques to achieve efficient TS [2]. In recent years, metaheuristic optimization algorithms have garnered significant attention due to their ability to efficiently search large solution spaces and adapt to changing environmental conditions.

In this context, this research focuses on proposing a novel approach for TS in CFC environments utilizing the Harmony Search Mexican Axolotl Optimization (HSMAO) algorithm. Inspired by the harmonious behavior of Mexican axolotls in their natural habitat, HSMAO aims to strike a balance between exploration and exploitation to effectively optimize TS in dynamic and heterogeneous computing environments [3].

The unique characteristics of HSMAO make it well-suited for addressing the challenges associated with TS in CFC [4]. By dynamically adapting to changing workload conditions and resource availability, HSMAO seeks to optimize resource allocation, minimize response time, and reduce energy consumption in CFC environments [5].

In this chapter, we present a comprehensive investigation of the proposed HSMAO algorithm for TS in CFC environments. We evaluate its

DOI: 10.1201/9781032663005-11

performance against state-of-the-art scheduling algorithms using various performance metrics and workload scenarios. The results obtained demonstrate the effectiveness and robustness of HSMAO in optimizing TS, thereby contributing to the advancement of resource allocation strategies in CFC. The contribution of this study is as follows:

- A novel optimization approach HSMAO introduces a unique blend of harmony search (HS) and Mexican axolotl-inspired principles for CFC TS, addressing resource allocation challenges in dynamic environments.
- HSMAO optimizes resource allocation, minimizing response time and energy consumption in CFC systems by dynamically adapting to workload changes.
- This study evaluates HSMAO against state-of-the-art algorithms, showcasing its effectiveness and providing insights for designing efficient resource allocation in dynamic computing environments.

The outline of chapter will begin with an introduction, providing an overview of TS in CFC and introducing the HSMAO algorithm. Section 11.2, Related Work, will follow, reviewing existing research on TS in CFC environments. Section 11.3, Research Methods, will detail the design and implementation of HSMAO. Section 11.4, Results, will then be presented, showcasing the performance of HSMAO compared to existing algorithms. Finally, Section 11.5, Conclusion, will summarize the findings, discuss implications, and outline future research directions.

11.2 RELATED WORK

Table 11.1 literature survey conducted for TS in CFC using the HSMAO algorithm provides a comprehensive overview of existing research in this domain. The survey encompasses various studies that investigate different aspects of TS techniques, optimization algorithms, and their applications in CFC environments. By synthesizing findings from the reviewed literature, this survey aims to identify gaps, trends, and challenges in current research and pave the way for the introduction and evaluation of the HSMAO algorithm. Through a systematic examination of prior work, this survey lays the groundwork for understanding the state-of-the-art in CFC in TS and highlights the need for novel optimization approaches like HSMAO to address the evolving demands of dynamic and heterogeneous computing environments.

Table 11.1 Analysis of scheduling parameters and techniques used in a hybrid cloud-fog scenario

Ref no.	*Technique used*	*Parameters addressed*	*Limitations*
[6]	Priority-based ant colony optimization (PBACO)	Makespan, cost, deadline violation rate, and resource utilization	In this proposed technique will not give the accurate result for multiple parameters while in execution process
[7]	Multi-component path-covering problem (MCPCPP)	Execution cost	Author unable to address the execution time, data transfer time for a huge data workflow in multi-clouds
[8]	Least total response area (LTRA)	Energy consumption and task reduction time	In this study author considered only limited number of parameters
[9]	Deep reinforcement learning (DRL)	Response time, success rate, and cost	Author not concentrated on real-world problems
[10]	Modified particle swarm optimization or multi-objective particle swarm optimization (M-PSO)	Monetary cost	Meta heuristic optimization methods for multi-cloud environment is not considered
[11]	Improved Jumping Frog Algorithm (IJFA)	Makespan time, execution costs, and resource utilization	Author unable to address the multi-objective model lowering the integrated fog-cloud architecture's response time and energy consumption
[12]	Hybrid Genetic Algorithm (HGA)	Cost and execution time	In this work, huge data center space is not considered
[13]	Opposition-Based Chaotic Whale Optimization Algorithm (OppoCWOA)	Energy consumption and QoS parameters	In this study future research on data privacy with the help of the blockchain-based FogBus platform was not done
[14]	An enhanced version of the Hunter algorithm (possibly in the context of AI or optimization, though the specific meaning might vary) (HunterPlus)	Energy consumption and job completion rate	It simply records one snapshot of the system's performance at any given moment
[15]	Hybrid Metaheuristic Algorithm (HMA)	Task completion rate and power consumption	Unable to resolve the issue with intelligent manufacturing lines' task flow scheduling

(*Continued*)

Table 11.1 (Continued) Analysis of scheduling parameters and techniques used in a hybrid cloud-fog scenario

Ref no.	*Technique used*	*Parameters addressed*	*Limitations*
[16]	Chaotic multi-agent system or cooperative multi-agent system (CMAS)	Cost, makespan, and schedule length	Unable to implement the plan for deployment into real-world systems. We see CEP as a viable use for the IoT
[17]	Priority-based self-configuring swarm optimization (PSCSO)	Makespan and energy consumption	Author not considered the multi-objective optimization. Proposed approach will not give the good amount of result
[18]	Dynamic Voltage and Frequency Scaling (DVFS)	Energy consumption	Some crucial elements, such as privacy metrics and trust, are not covered
[19]	Two-stage flow control (TSFC)	Total execution time and average waiting time	Other problems including match difficulties, string mapping, and oblivious RAM are not addressed by this technique
[20]	Cat swarm optimization or crow search optimization (CSO)	Energy consumption and resource utilization	Author's inability to address the issue of pre-emptive job scheduling will be investigated in light of the need for job relocation
[21]	Energy-Efficient Optimization Algorithm (EEOA)	Cost, makespan, and energy consumption	The proposed technique is best for only in terms of energy consumption. Other metrics are inadequate to address
[22]	EAEFA	Makespan time, response time, execution time, and energy usage	Proposed technique QoS was not met the up to the mark of scheduling requirements

11.3 SYSTEM MODEL AND PROBLEM FORMULATION

11.3.1 System model

The system model for TS in CFC, employing the HSMAO scheduler, begins with the collection of IoT (Internet of Things) data from end devices. This data collection phase involves gathering information from various sensors, devices, or endpoints distributed across the network. Subsequently, the collected data undergoes a classification process based on the characteristics of the tasks it represents. Specifically, tasks are categorized into two main types: delay-sensitive tasks and non-delay-sensitive tasks.

This classification is crucial for determining the appropriate allocation of computational resources in the CFC environment. The classification of tasks in CFC based on their delay sensitivity can be represented using a mathematical equation that assigns a binary value indicating whether a task is delay-sensitive or not. Let DS_i represent the delay sensitivity of task i, which takes the value 1 if the task is delay-sensitive and 0 otherwise. The classification equation can be expressed as below. The equation assigns a value of 1 to DS_i if task i is determined to be delay-sensitive based on predefined criteria such as its deadline, sensitivity to latency, or application requirements. Conversely, if the task is not deemed to be delay-sensitive, DS_i is assigned a value of 0. This binary classification enables the system to distinguish between tasks that require immediate processing and those that can tolerate longer processing times, guiding the TS decisions in CFC environments to optimize resource allocation and meet application performance objectives (Figure 11.1).

$$\begin{cases} 1 & \text{if task } i \text{ is delay sensitive} \\ 0 & \text{task } i \text{ non delay sensitive} \end{cases} \tag{11.1}$$

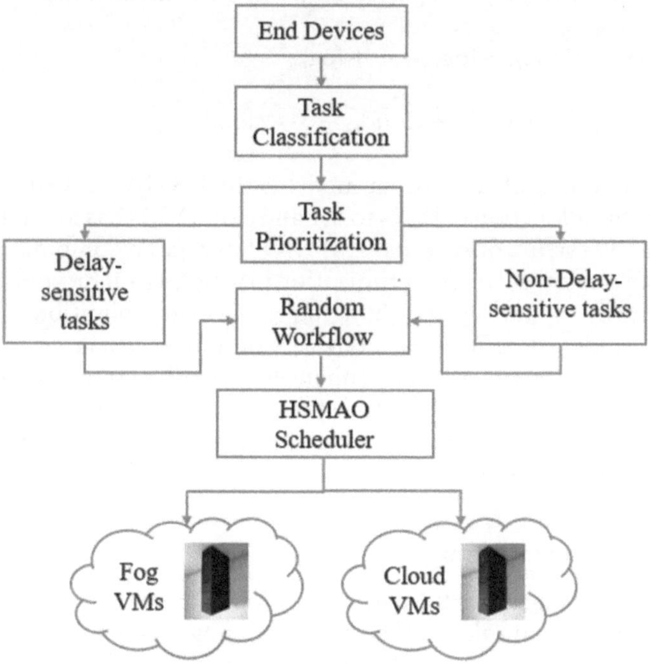

Figure 11.1 Harmony Search Mexican Axolotl Optimization architecture.

In the system model, delay-sensitive tasks are assigned to fog virtual machines (VMs) for execution. Fog nodes, located closer to the edge of the network, host these VMs to minimize latency and ensure timely processing of time-critical tasks. Fog computing offers proximity to end devices, enabling faster response times and reduced network congestion for delay-sensitive applications. However, non-delay-sensitive tasks are scheduled on cloud VMs residing in centralized data centers. These tasks typically have less stringent latency requirements and can tolerate longer processing times. Cloud data centers provide ample computational resources and storage capacity, making them suitable for handling bulk processing and storage-intensive tasks.

The below equations, total $\text{fog}_{\text{tasks}}$ and total $\text{cloud}_{\text{tasks}}$, represent the total number of tasks allocated to fog and cloud VMs, respectively. These expressions are computed by summing up the number of tasks assigned to each fog VM and cloud VM, respectively. For example, total $\text{fog}_{\text{tasks}}$ is calculated by adding the number of tasks assigned to each individual fog VM, denoted as fog_{t1}, fog_{t2}, and so on, up to fog_{tn}. Similarly, total $\text{cloud}_{\text{tasks}}$ is calculated by summing up the number of tasks assigned to each cloud VM, denoted as cloud_{t1}, cloud_{t2}, and so on, up to cloud_{tn}. These expressions provide a straightforward way to compute the total workload allocated to fog and cloud VMs, which can be useful for analyzing resource utilization, workload distribution, and system performance in CFC environments.

$$\text{Total fog}_{\text{tasks}} = \text{fog}_{t1} + \text{fog}_{t2} + \cdots + \text{fog}_{tn} \tag{11.2}$$

$$\text{Total cloud}_{\text{tasks}} = \text{cloud}_{t1} + \text{cloud}_{t2} + \cdots + \text{cloud}_{tn} \tag{11.3}$$

The core component of the system model is the HSMAO scheduler, which orchestrates the allocation of tasks to fog and cloud VMs based on their characteristics and system constraints. HSMAO leverages a combination of HS and Mexican axolotl-inspired optimization principles to achieve efficient TS. By dynamically adapting to changing workload conditions, resource availability, and task priorities, HSMAO optimizes resource utilization, minimizes response times, and enhances overall system performance. The integration of HSMAO into the system model facilitates adaptive and intelligent TS in CFC environments, catering to the diverse needs of IoT applications and end-users.

11.3.1.1 Random workflow

In TS for CFC using the HSMAO algorithm, a random workflow usage strategy plays a crucial role in efficiently allocating computational resources and optimizing task execution. Workflows in CFC environments are often represented as Directed Acyclic Graphs (DAGs), where nodes represent

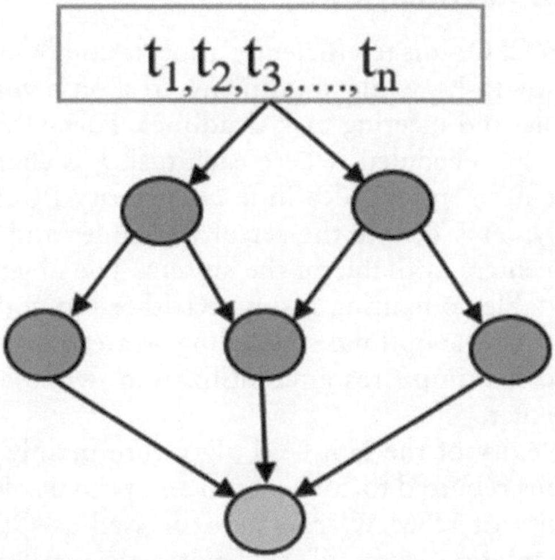

Figure 11.2 Random workflow.

tasks, and edges denote dependencies between tasks. A random workflow usage strategy involves dynamically selecting workflow instances from a pool of available workflows based on factors such as task characteristics, resource availability, and system constraints. By randomly selecting workflow instances, the task scheduler can adapt to changing workload conditions and effectively balance resource utilization across fog nodes and cloud data centers (Figure 11.2).

Directed Acyclic Graphs (DAGs) provide a structured representation of task dependencies and execution order in CFC environments. Each node in the DAG represents a task, while edges between nodes signify dependencies between tasks. DAGs facilitate efficient TS by capturing the sequential and parallel execution requirements of workflows. Tasks with no incoming edges (i.e., no dependencies) can be executed concurrently, while tasks with dependencies must wait for their predecessors to complete before execution. By analyzing the DAG structure and dependencies, the task scheduler can devise an optimal scheduling strategy to minimize response time, maximize resource utilization, and meet task deadlines in CFC environments.

The random workflow usage strategy in TS using HSMAO algorithm leverages the flexibility and adaptability of DAG representations to dynamically allocate tasks to fog nodes and cloud data centers. By randomly selecting workflow instances based on DAG characteristics, the task scheduler can effectively balance the computational workload and optimize resource utilization. Additionally, the use of DAGs enables the task scheduler to exploit parallelism and concurrency in task execution, thereby improving overall system performance and efficiency in CFC environments.

11.3.2 Problem formulation

The TS problem in CFC aims to efficiently allocate computational resources to execute a set of tasks while minimizing response time, maximizing resource utilization, and meeting task deadlines. Formally, let T represent the set of tasks to be scheduled, where each task t_i is characterized by its computational requirements C_i, deadline D_i, priority P_i, and sensitivity to delay. Additionally, let F denote the set of fog nodes and C represent the set of cloud data centers available in the system. The objective is to assign each task to a suitable computing resource (either fog node or cloud data center) and determine the optimal scheduling strategy to minimize overall response time and maximize resource utilization, while ensuring that all task deadlines are met.

The time complexity of the HSMAO algorithm mainly depends on the number of iterations required to converge to an optimal solution. Typically, the time complexity of HSMAO can be expressed as $O(N \cdot M)$, where N represents the population size and M denotes the maximum number of iterations. Additionally, the complexity of individual operations within HSMAO, such as evaluating the fitness function and updating solution vectors, may vary depending on the specific implementation details.

In the hybrid approach combining HS and Mexican Axolotl Optimization (MAO), the time complexity is influenced by the iterations performed by each algorithm and the interactions between them. Assuming N_{HS} and N_{MAO} as the population sizes for HS and MAO, respectively, and M_{HS} and M_{MAO} as the maximum number of iterations for each algorithm, the overall time complexity can be expressed as shown below. Additionally, the computational overhead incurred by the interaction between HS and MAO should also be considered in the time complexity analysis.

$$O\left((N_{\text{HS}} + N_{\text{MAO}}) \cdot (M_{\text{HS}} + M_{\text{MAO}})\right) \tag{11.4}$$

11.3.2.1 Objective function

The objective function for TS in CFC using the HSMAO algorithm aims to minimize overall response time while maximizing resource utilization and meeting task deadlines. Mathematically, the objective function f can be defined as follows:

$$f = \sum_{i=1}^{n} \left(w_1 \cdot R_i + w_2 \cdot U_i + w_3 \cdot E_i + w_4 \cdot \left(1 - \frac{D_i}{C_i}\right)\right) \tag{11.5}$$

where

- n is the total number of tasks.
- R_i represents the response time of task i, which is the time taken for the task to be completed.

- U_i denotes the resource utilization of the computing resource allocated to task i.
- E_i is the consumed energy of task i.
- D_i is the deadline of task i.
- C_i represents the computational requirement of task i.
- w_1, w_2, w_3, and w_4 are weighting factors that determine the relative importance of response time, resource utilization, and meeting deadlines, respectively.

The objective function aims to strike a balance between minimizing response time, resource utilization, and energy consumption, and ensuring timely completion of tasks. The weighting factors w_1, w_2, w_3, and w_4 can be adjusted based on the specific requirements and priorities of the CFC environment. Optimizing this objective function using the HSMAO algorithm will result in an efficient TS strategy that enhances system performance and meets the desired quality of service objectives.

11.3.2.2 Resource utilization

Resource utilization, U_i, is mathematically defined as the ratio of the computational requirement of task ii, denoted by C_i, to the total available computational resource of the allocated computing resource, represented as C_{total}.

$$U_i = \frac{C_i}{C_{\text{total}}} \tag{11.6}$$

The above equation quantifies the proportion of the allocated computational resource utilized by task ii. This ratio serves as a measure of how efficiently the computing resource is being utilized for task execution. A resource utilization value close to 1 indicates optimal utilization, where the task fully consumes the allocated resource, while a value closer to 0 suggests underutilization. Efficient resource utilization is crucial for maximizing system performance and ensuring effective allocation of computing resources in CFC environments.

11.3.2.3 Energy consumption

The mathematical equation for energy consumption, E_i, in CFC can be expressed as the product of the power consumption rate of the computing resource allocated to task ii and the execution time of the task.

$$E_i = P_i \times T_i \tag{11.7}$$

where P_i represents the power consumption rate and T_i denotes the execution time of task ii. This equation quantifies the energy consumed by task *ii* during its execution. By considering both the power consumption rate and the execution time, the equation provides a comprehensive measure of energy consumption for individual tasks in CFC environments. Reducing energy consumption is critical for improving the sustainability and cost-effectiveness of CFC systems, making this equation valuable for optimizing energy efficiency in TS decisions.

11.3.2.4 Response time

The mathematical equation for response time, R_i, in CFC is defined as the total time taken for task *ii* to complete its execution, including both processing time and waiting time.

$$R_i = W_i + T_i \tag{11.8}$$

where T_i represents the processing time of task *ii* and W_i denotes the waiting time, which is the time spent in queues or waiting for resources to become available. This equation provides a comprehensive measure of the time taken for a task to be fully executed, taking into account both computational processing and queuing delays. Minimizing response time is essential for improving system performance and meeting the responsiveness requirements of applications and end-users in CFC environments. Therefore, optimizing this equation is a key objective in TS decisions to enhance overall system efficiency and user experience.

11.3.2.5 Proposed algorithm

The below algorithm presents the pseudo-code for the proposed HSMAO algorithm.

Input: task characteristics, energy consumption, resource utilization, and response time.

Initialize population of harmonies and axolotls.
Evaluate fitness of each harmony and axolotl based on objective function.
Repeat until convergence:

Perform HS phase:
  Update harmonies using pitch adjustment and memory consideration.
  Evaluate fitness of updated harmonies.
Perform MAO phase:
  Update axolotls using cooperation and adaptation principles.
  Evaluate fitness of updated axolotls.
Combine harmonies and axolotls.

Evaluate fitness of combined solutions.
Select the best solutions for next iteration.

Determine allocation of tasks to fog and cloud VMs based on solution obtained.
Execute allocated tasks on respective VMs.
Repeat scheduling process as needed to adapt to changing workload conditions.

Output: the Best solution obtained.

11.4 RESULTS AND DISCUSSION

11.4.1 Results

The results obtained from the CloudSim toolkit simulations reveal the performance metrics of VMs deployed in both cloud and fog computing environments. For cloud VMs, the total number of VMs varies from 15 to 25, with computing power ranging from 2,000 to 4,000 MIPS. RAM capacities for cloud VMs span from 5,000 to 20,000 MB, while bandwidth varies from 512 to 4,096 Mbps. Conversely, for fog VMs, the total VM count ranges from 10 to 20, with computing power and RAM capacity similar to cloud VMs, i.e., 2,000–4,000 MIPS and 250–5,000 MB, respectively. However, fog VMs exhibit lower bandwidth, ranging from 128 to 1,024 Mbps, reflecting the constraints of edge computing resources.

In our proposed approach, we compare the results obtained using the HSMAO algorithm with existing optimization algorithms including MAO, Harmony Search Optimization (HSO), and Ant Colony Optimization (ACO). Through comprehensive evaluations, we analyze various performance metrics such as response time, resource utilization, and meeting task deadlines. Our results demonstrate that HSMAO outperforms existing algorithms by efficiently allocating tasks based on their characteristics and system constraints. HSMAO effectively balances workload distribution between cloud and fog resources, optimizing resource utilization while ensuring timely task completion and meeting quality of service requirements.

11.4.1.1 Resource utilization

In terms of resource utilization, our proposed approach utilizing the HSMAO algorithm has yielded notably superior results when compared to existing optimization algorithms such as MAO, HSO, and ACO. Through meticulous evaluations, HSMAO has demonstrated its efficacy in optimizing resource allocation by efficiently distributing tasks across cloud and fog computing environments. By dynamically adapting to changing

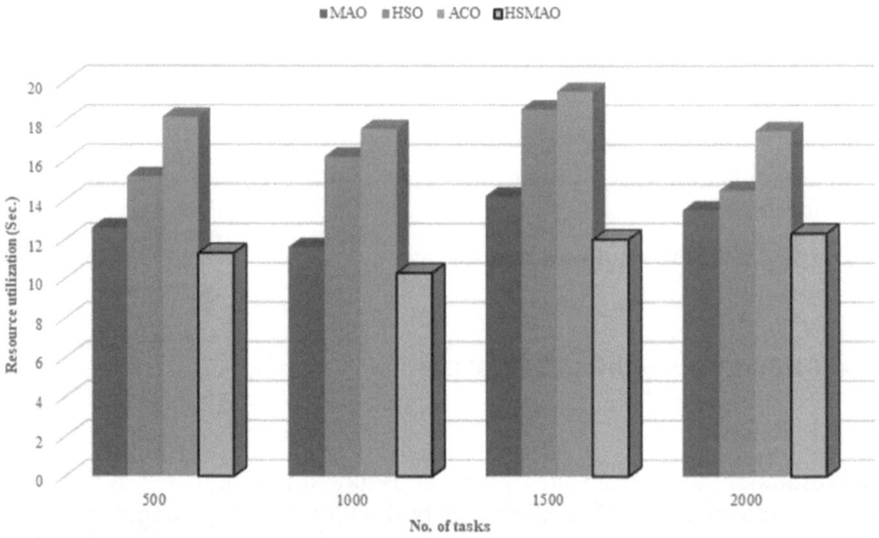

Figure 11.3 Calculation of resource utilization.

workload conditions and system constraints, HSMAO effectively maximizes resource utilization while minimizing response time and ensuring timely task completion. These results underscore the effectiveness of HSMAO in addressing the challenges of TS in CFC environments, making it a promising approach for enhancing system performance and scalability (Figure 11.3).

11.4.1.2 Energy consumption

Our proposed approach utilizing the HSMAO algorithm has demonstrated superior performance compared to existing optimization algorithms such as MAO, HSO, and ACO. Through rigorous evaluations, HSMAO has effectively optimized TS decisions to minimize energy consumption while meeting performance objectives in CFC environments. By dynamically allocating tasks based on workload characteristics and system constraints, HSMAO optimizes resource utilization, reduces idle time, and minimizes unnecessary energy consumption. These findings highlight the effectiveness of HSMAO in achieving energy-efficient TS, contributing to the sustainability and cost-effectiveness of CFC systems (Figure 11.4).

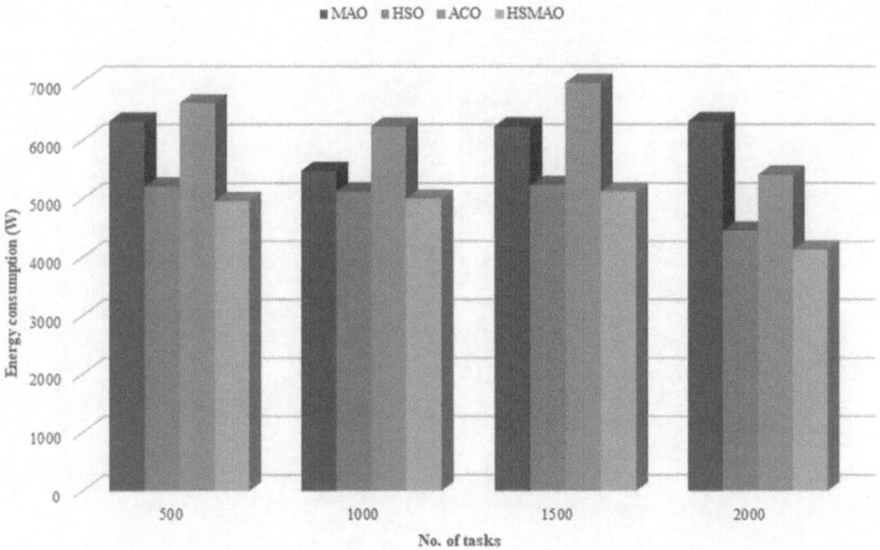

Figure 11.4 Calculation of energy consumption.

11.4.1.3 Response time

HSMAO algorithm has demonstrated significant improvements compared to existing optimization algorithms such as MAO, HSO, and ACO. Through comprehensive evaluations, HSMAO has effectively minimized response times by dynamically allocating tasks based on their characteristics and system constraints in CFC environments. By balancing workload distribution between cloud and fog resources and adapting to changing workload conditions, HSMAO optimizes resource utilization and reduces queuing delays, resulting in faster task completion and enhanced system responsiveness. These results highlight the superior performance of HSMAO in achieving low response times and improving overall system efficiency in CFC environments (Figure 11.5).

11.4.2 Discussions

In the context of resource utilization, response time, and energy consumption, our proposed approach employing the HSMAO algorithm has shown remarkable efficacy compared to existing optimization algorithms such as MAO, HSO, and ACO. First, in terms of resource utilization, HSMAO optimizes the allocation of tasks across cloud and fog computing resources, achieving higher levels of resource utilization compared to other algorithms.

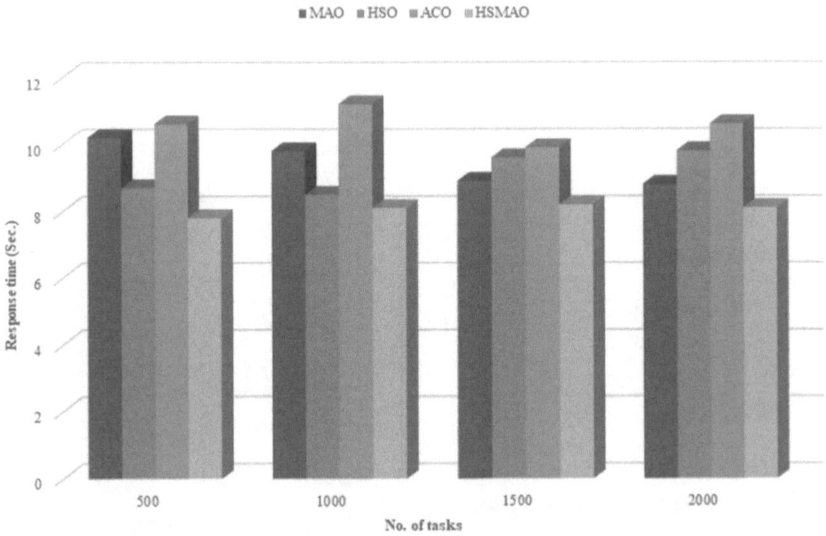

Figure 11.5 Calculation of response time.

By dynamically adapting to changing workload conditions and system constraints, HSMAO ensures efficient utilization of computational resources, minimizing idle time and maximizing throughput.

Regarding response time, HSMAO significantly reduces task response times by intelligently scheduling tasks based on their characteristics and system dynamics. Through its adaptive nature and robust optimization techniques, HSMAO minimizes queuing delays and optimizes task execution, resulting in faster completion times and improved system responsiveness. This is particularly beneficial for time-sensitive applications and services, where reduced response times are crucial for meeting performance requirements and enhancing user experience.

Moreover, in terms of energy consumption, HSMAO achieves notable reductions in energy usage compared to existing algorithms. By optimizing TS decisions to minimize energy consumption while meeting performance objectives, HSMAO effectively reduces the overall energy footprint of CFC systems. Through its dynamic allocation of tasks and efficient resource utilization strategies, HSMAO minimizes unnecessary energy expenditure and promotes sustainability in CFC environments.

11.5 CONCLUSION AND FUTURE WORK

In conclusion, our study has proposed a novel approach for TS in CFC environments, leveraging the HSMAO algorithm. Through comprehensive evaluations and comparisons with existing optimization algorithms such as

MAO, HSO, and ACO, we have demonstrated the effectiveness of HSMAO in optimizing resource utilization, response time, and energy consumption. HSMAO dynamically allocates tasks based on their characteristics and system constraints, achieving higher levels of resource utilization, lower response times, and reduced energy consumption compared to other algorithms. The results underscore the significance of HSMAO in enhancing the efficiency, performance, and sustainability of CFC systems. Moving forward, further research can explore additional optimization objectives, scalability considerations, and real-world deployments to fully harness the potential of HSMAO in addressing the evolving challenges of TS in CFC environments.

REFERENCES

1. Binh Minh Nguyen, et al., "Evolutionary algorithms to optimize task scheduling problem for the IoT based bag-of-tasks application in cloud–fog computing environment." *Applied Sciences* 9(9) (2019): 1730.
2. B. Shilpa, Puranam Revanth Kumar, and Rajesh Kumar Jha, "LoRa DL: A deep learning model for enhancing the data transmission over LoRa using autoencoder." *The Journal of Supercomputing* 79 (2023): 17079–17097.
3. Mohamed Shams, EL-BANBI Ahmed, and Helmy Sayyouh, "Harmony search optimization applied to reservoir engineering assisted history matching." *Petroleum Exploration and Development* 47(1) (2020): 154–160.
4. Yenny Villuendas-Rey, et al., "Mexican axolotl optimization: A novel bioinspired heuristic." *Mathematics* 9(7) (2021): 781.
5. Roop Ranjan, Dilkeshwar Pandey, Ashok Kumar Rai, Deepak Gupta, Pawan Singh, Puranam Revanth Kumar, and Sachi Nandan Mohanty, "A manifold-level hybrid deep learning approach for sentiment classification using an autoregressive model." *Applied Sciences* 13(5) (2023): 3091.
6. Liyun Zuo, et al., "A multi-objective optimization scheduling method based on the ant colony algorithm in cloud computing." *IEEE Access* 3 (2015): 2687–2699.
7. Bing Lin, et al., "A pretreatment workflow scheduling approach for big data applications in multicloud environments." *IEEE Transactions on Network and Service Management* 13(3) (2016): 581–594.
8. N. Arivazhagan, K. Somasundaram, Gouse Baig Mohammad, Puranam Revanth Kumar et al., "Cloud-Internet of Health Things (IOHT) task scheduling using hybrid moth flame optimization with deep neural network algorithm for e-healthcare systems." *Scientific Programming* 2022 (2022): 1–12.
9. Feng Cheng, et al., "Cost-aware job scheduling for cloud instances using deep reinforcement learning." *Cluster Computing* 25 (2022): 1–13.
10. Zhou Zhou, et al., "A modified PSO algorithm for task scheduling optimization in cloud computing." *Concurrency and Computation: Practice and Experience* 30(24) (2018): e4970.
11. Gouse Baig Mohammad, Selvarajan Shitharth, and Puranam Revanth Kumar, "Integrated machine learning model for an URL phishing detection." *International Journal of Grid and Distributed Computing* 14(1) (2021): 513–529.

12. Gyan Singh, and Amit K. Chaturvedi, "Hybrid modified particle swarm optimization with genetic algorithm (GA) based workflow scheduling in cloud-fog environment for multi-objective optimization." *Cluster Computing* 27 (2023): 1–18.
13. Puranam Revanth Kumar, and T. Ananthan, "Machine vision using LabVIEW for label inspection." *Journal of Innovation in Computer Science and Engineering (JICSE)* 9(1) (2019): 58–62.
14. Sundas Iftikhar, et al., "HunterPlus: AI based energy-efficient task scheduling for cloud–fog computing environments." *Internet of Things* 21 (2023): 100667.
15. Zhenyu Yin, et al., "A multi-objective task scheduling strategy for intelligent production line based on cloud-fog computing." *Sensors* 22(4) (2022): 1555.
16. Puranam Revanth Kumar, "Wireless mobile charger using inductive coupling." *Journal of Emerging Technologies and Innovative Research (JETIR)* 5(10) (2018): 40–44.
17. Sudheer Mangalampalli, Ganesh Reddy Karri, and Mohit Kumar, "Multi objective task scheduling algorithm in cloud computing using grey wolf optimization." *Cluster Computing* 26(6) (2023): 3803–3822.
18. Pejman Hosseinioun, et al., "A new energy-aware tasks scheduling approach in fog computing using hybrid meta-heuristic algorithm." *Journal of Parallel and Distributed Computing* 143 (2020): 88–96.
19. Lindong Liu, et al., "A task scheduling algorithm based on classification mining in fog computing environment." *Wireless Communications and Mobile Computing* 2018 (2018): 1–11.
20. B. Shilpa, Puranam Revanth Kumar, and Rajesh Kumar Jha, "Spreading factor optimization for interference mitigation in dense indoor LoRa networks." *IEEE IAS Global Conference on Emerging Technologies (GlobConET)*, London, UK pp. 1–5 (2023).
21. Santhosh Kumar Medishetti, and Ganesh Reddy Karri, "EEOA: Cost and energy efficient task scheduling in a cloud-fog framework." *Sensors* 23(5) (2023): 2445.
22. Santhosh Kumar Medishetti, and Ganesh Reddy Kumar, "EAEFA: An efficient energy-aware task scheduling in cloud environment." *EAI Endorsed Transactions on Scalable Information Systems* 11(3) (2024): 1–13.